The Open University

Arts/Mathematics/Science/Technology
An Inter-faculty Second Level Course in the History of Science

Science and Belief: from Copernicus to Darwin
Block V Units 12–14

The Crisis of Evolution

The Open University Press

D. Daiches Raphael

1961 - J Dillenberger
"Protestant Thought and
Natural Science"
Collins, 1961

Studies in the Hist + Phil. of Sci
vol II = 1971

COVER ILLUSTRATION:

Left: T. H. Huxley, Vanity Fair *cartoon by 'Ape', 28 January 1871; centre: Charles Darwin,* Vanity Fair *cartoon by 'Ape', 30 September 1871; right: Bishop Samuel Wilberforce,* Vanity Fair *cartoon by 'Ape', 24 July 1869 (Mary Evans Picture Library and Mansell Collection).*

The Open University Press,
Walton Hall, Milton Keynes.
MK7 6AA

First published 1974. Reprinted 1975.

Designed by the Media Development Group of the Open University.

Printed in Great Britain by
MARTIN CADBURY, A SPECIALIZED DIVISION OF SANTYPE INTERNATIONAL,
WORCESTER AND LONDON.

ISBN 0 335 00785 6

This text forms part of an Open University course. The complete list of units in the course appears at the end of this text.

For general availability of supporting material referred to in this text please write to the Director of Marketing, The Open University, P.O. Box 81, Walton Hall, Milton Keynes, MK7 6AT.

Further information on Open University courses may be obtained from the Admissions Office, The Open University, P.O. Box 48, Walton Hall, Milton Keynes, MK7 6AB.

1.2

Contents

Set Books

C. C. Gillispie, *Genesis and Geology*, Harper and Row.

D. C. Goodman (ed.), *Science and Religious Belief: A Selection of Primary Sources*, John Wright and Sons/The Open University Press (ANTHOLOGY I).

R. Hooykaas, *Religion and the Rise of Modern Science*, Scottish Academic Press/Chatto and Windus.

C. A. Russell (ed.), *Science and Religious Belief: A Selection of Recent Historical Studies*, University of London Press/The Open University Press (ANTHOLOGY II).

Unit 12
Precursors of Darwin?

Prepared by John Hedley Brooke for the Course Team

ILLUSTRATION OVERLEAF

Portraits of Lamarck (top left), Buffon (top right), Lyell (bottom left) and Cuvier (bottom right) (Mansell Collection).

Contents

1 Introduction

The object of this unit is to help you to understand the background of the nineteenth-century debate concerning the meaning and implications of the Darwinian theory of evolution. It is not about the theory itself, or indeed *primarily* about the theories held by Darwin's 'precursors'. It is more directly concerned with the problems which confront the historian when he turns his attention to so difficult a subject. Instead of pursuing a systematic analysis of the thought of pre-Darwinian naturalists, we shall ask more elementary questions about existing analyses and their limitations. Instead of parading Darwin's precursors we shall try to answer the ulterior question whether it is meaningful to see them as precursors at all. It is not so very long since books could be published with titles like *Forerunners of Darwin*, or *Just before Darwin*, which implied that the pre-Darwinian period could best be approached by starting with Darwin and working backwards. Critical scholarship in the history of biology has increasingly revolted against this assumption, until we now have a situation where historians are likely to be chastised if they use the word 'evolution' at all when discussing Darwin's predecessors.

One reason for this is that the word 'evolution' in the late eighteenth and early nineteenth centuries did not mean what it means to us. It was almost always used to mean the development (the 'evolution') of a foetus, i.e. its proper application was in the context of embryology. It could equally be used to refer to theories of dynamical astronomy—theories like that of Laplace, which displaced static models of the universe with others which presupposed the emergence of the present configuration of the heavens over a very long period of time. Oddly enough it was a word that Darwin himself rarely used. The danger should then be obvious: the word is charged with connotations which it has not always had. If we speak uncritically of 'evolution before Darwin' we shall run the risk of imposing neo-Darwinian strictures and structures on the ideas of an earlier age. Does that mean that nobody before Darwin did believe in the progressive transformation of species? The answer is a definite 'no'. There were those who did, and you will be meeting them in due course. We shall from time to time refer to them as 'evolutionists' but you must remember that about the only thing they had in common was a belief that the animals which roam the globe today have been derived from more primitive ones. When the word 'evolution' appears in the text it will mean nothing more than that, nothing more than the transformation of living things over the course of time. With that qualification it does not become irresponsible to discuss 'evolutionary speculation' before Darwin. Whether it is constructive to do so is one of the things you will have to decide for yourself.

Objectives

After working through this unit you should be able to:

1 Enumerate conceptual problems which had a bearing on the fortunes of evolutionary speculation, and explain how each one could be associated with aspects of religious belief.
2 Discuss the limitations of the historiographical tradition which digs for 'precursors'; and of the 'warfare' tradition in its application to Buffon and Cuvier.
3 Expound the main features of Lamarck's development hypothesis, and the reasons why Lyell rejected it.
4 Evaluate the significance of Chambers' *Vestiges of the Natural History of Creation* (1844), and explain why it was so controversial a book.
5 State the case for regarding the natural theologians of the Anglican Church as those to whom Darwin was most indebted; examine its strengths and weaknesses; and explain the relevance of Malthus to the thesis in question.

2 The Biological Background

Since this is our first detailed excursion into the history of biology we shall begin the unit by locating a cluster of problems which may help you to appreciate the scientific and philosophical background to the evolutionary debates. These problems were frequently interrelated and they all had some bearing on the course of evolutionary speculation. Nor could they be discussed without reference to the religious preconceptions which helped to determine what constituted acceptable solutions. As Maurice Mandelbaum stressed when delineating the scientific background of evolutionary theory: 'when the specific problems which absorbed men of science were discussed, these problems were regarded as being of religious import, and were (more often than not) explicitly discussed in these terms'.[1] The problems that we have chosen to consider are by no means exhaustive, but they should help you to see that opposition to evolutionary speculation was frequently based on the scientific knowledge of the day. Precisely because of that, a word or two of warning might not be out of place at the start. When we draw attention to arguments and ideas which made it difficult to accept the transformation of species, we are not passing judgement—not implying that this was perforce a bad thing. It would, in fact, be an absurd assumption to suppose that Darwin's predecessors *ought* to have seen the truth of 'evolution' if only they had got their facts straight! Nor do we intend to suggest that a theory of evolution was bound to emerge from a study of these particular problems. There was no inevitable march towards Darwin.

2.1 The nature of life

Perhaps the most fundamental questions raised by the study of living things concern the nature of life itself. What are the primary characteristics of systems which exhibit life? Is there a clear dividing line between the animate and the inanimate? Do other animals experience consciousness as we do? Is there something ultimately mysterious, some irreducible spark of life, which distinguishes the living from the dead? These are grandiose questions and each one doubtless suggests a dozen more! To discuss any one of them in depth would be a vast undertaking, but historians have been able to simplify the picture by distinguishing between two recurring attitudes towards the investigation of living things. One is usually called 'mechanistic'; the other is 'vitalistic'. There are grave problems of definition with these terms but to a first approximation a 'mechanist' may be said to be one who believes that in the long run, if not at present, the basic concepts of biology may be reduced to those of the physical sciences: ultimately there will be no mystery: life is just a bag of self-organizing chemicals obeying strict physical laws. The vitalist, on the contrary, is one who rejects the reductionist thesis and who maintains that no matter how sophisticated the physical sciences may become, there will always remain some features of organic phenomena which can only be explained by postulating some inherent organizing principle, power or force. Vitalists have of course differed among themselves as to what these features might be, but perhaps you could suggest some of them for yourself.

If you were a vitalist what remarkable properties of living things might you appeal to?

My own list would include such items as the apparently purposeful activity of living creatures, their instincts, their ability to move themselves and to reproduce; and in the case of man, his self-consciousness. Within an organism one could point to the maintenance of a constant temperature, the highly selective processes of absorption and secretion, and the power of the body to heal itself.

[1] M. Mandelbaum, 'Scientific Background of Evolutionary Theory in Biology', in P. Wiener and A. Noland (eds.), *Roots of Scientific Thought*, Basic Books, New York, 1957, p. 517.

Some of these features were of course more mysterious in previous centuries than they are in our own, and for that reason biologists have sometimes written the history of their subject as a long but steady triumph of mechanism over vitalism—the march led by Harvey, the discoverer of the circulation of the blood and reaching its destination in Darwin who dispelled the mysteries of speciation. But this is far too crude an approach—for several reasons. Insofar as the labels 'mechanism' and 'vitalism' are useful at all it is more accurate to see the repeated swing of the pendulum from the one to the other over the last four centuries than it is to see a linear, progressive, defeat of vitalism. As you saw in Units 4 and 5, Descartes established a programme which required that physiologists should proceed with the assumption that animals were nothing but complex machines. But during the eighteenth century the Cartesian programme had been found wanting, and alternative vitalist schools flourished again for a while. It had proved impossible to find convincing mechanical analogues for those complicated and selective phenomena which implied the activity of some intelligent agent within an organism. The 'linear progress' approach is also crude because it rarely does justice to the position of the great biologists of the past. William Harvey has often been portrayed as a mechanist since it was he who discovered the circulation of the blood, comparing the action of the heart with that of a pump. But this is a serious oversimplification since, compared with Descartes, Harvey's emphasis on final causes was still Aristotelian, and, compared with Aristotle himself, Harvey was an emphatic 'vitalist' since he could repudiate the notion that the blood was merely composed of the four elements: there must, he argued, be an additional vital spirit. Those who like to applaud when vitalists are embarrassed, often miss the point that though many vitalist theories have indeed proved sterile, others in their day have been perfectly 'scientific'—as when Liebig in his *Animal Chemistry* (1842) insisted that his vital force obeyed the same kind of laws as the other forces in the scientists' repertoire. Finally, one ought not to overlook the fact that concepts introduced and retained in the life sciences do include some that would look out of place in the physical sciences. A good example would be the internal medium (*milieu intérieur*) of Claude Bernard.

These issues are so complex, and the changing status of a vitalist commitment so important for an understanding of the history of science that we are devoting a radio programme to this whole subject (Radio Programme 8). The problems are particularly fascinating in the context of scientific and religious beliefs, since a vitalist belief does not obviously belong to either category. It is *tempting* to suppose a strong affinity between vitalist and religious beliefs, not least because most religious orthodoxies have rejected the proposition that man is nothing but a complex machine. Certainly the historian would have no difficulty in finding examples of vitalist beliefs endorsed by a religious commitment. When John Ray (see Units 9–10) stressed that it was impossible for man to create living things he did so from a standpoint that was both vitalist and Christian. You will perhaps recall that he was committed to the vitalist, 'plastic', principle of Cudworth, and at the same time to a God who was so transcendent that His creative powers could never be matched. Again, a century later, certain religious beliefs of the Romantic poets were associated with their disdain for reductionist science: Coleridge could reproach those who, with the tools of the chemist, presumed to penetrate the 'sacred adyta' of organic life. But it would be dangerous to interpret the history of biology with the presupposition that vitalist and religious persuasions have always gone hand in hand or that mechanistic and religious philosophies have always been in opposition.

Can you say, from your knowledge of Descartes (see Unit 4), how his philosophy might constitute a counter-example to such presuppositions?

With the exception of the soul, Descartes' universe was entirely mechanistic. In other words, a religious commitment to the immortality of the human soul could be in holy alliance with an essentially mechanistic philosophy of life.

Another counter-example would be that of David Hume (see Units 9–10) who in his critique of natural theology had to find an explanation for the source of order and apparent design in the universe. As an alternative to the God of the physico-theologians Hume postulated a vitalistic principle, inherent within the universe, which might be equally capable of organizing matter into its present state. Vitalist and religious explanations could therefore be set up in opposition. It has even been argued by David Knight that one of the reasons why natural theology could become as mechanical in orientation as it did in the works of Paley, was a reaction against vitalist models of the universe which appeared to be derivatives of pantheism or other religious heterodoxies. The watch-world of Paley became the watch-word of natural theology as the vitalist alternatives were rebuffed.

So much for vitalism and religious beliefs, but what of vitalism and 'evolution'? Once again the issues are complicated, but there is no doubt that beliefs about the ultimate nature of life did have a bearing on evolutionary speculation. Erasmus Darwin who wrote poetry to celebrate the 'loves of the plants' (1789) was so convinced that plants were analogous to animals that he saw no reason why the first insects should not have developed from the stigmas of flowers! There is also a sense in which his theory of evolution was mechanistic, since, as a student of David Hartley's works on psychology, he believed that the male imagination could impress vibrations on an incipient embryo, thereby generating an offspring that would conform to the mental image of the parent. Those who could share Erasmus Darwin's belief that the lower animals were not without reason, saw in his theory a justification for the idea of progress: the lowliest creatures, visualizing a better state for their progeny at the moment of conception, would thereby generate more advanced forms.

Figure 1 Erasmus Darwin, portrait by Joseph Wright (National Portrait Gallery).

By the time that Erasmus Darwin's grandson, Charles, was publishing his theory of evolution, such speculations must have seemed pretty fanciful, but among those who are generally included as 'precursors' of Darwin, there were countless curious beliefs, some of them mechanistic, some more vitalistic. Since machines do not evolve or adapt by themselves, a completely mechanical view of nature could hardly have encouraged evolutionary speculation at all. Accordingly those who discussed the development of life within a deist framework (see Unit 11) usually required supplementary hypotheses of a vitalist kind. The French zoologist Lamarck provides perhaps the most famous example of this. You will have the opportunity to study his ideas in greater depth later in the unit, but basically he believed that the driving force behind evolution was a natural tendency towards a greater complexity, inherent within each creature. Though it may be unhelpful to classify Lamarck himself as either mechanist or vitalist, his interpreters often asserted that such a tendency towards greater complexity could not be explained in mechanistic terms alone.

Nor did it solve the problem when Robert Chambers, both anonymously and notoriously, suggested that the development of life was simply a consequence of a law that God had imposed upon Creation (*Vestiges*, 1844). By contrast, Charles Darwin's theory of evolution, despite its use of the troublesome metaphor 'natural selection', was to make no reference to irreducible vital forces or tendencies, and this is indoubtedly one of the reasons why it proved so acceptable in the long run. To many of Darwin's contemporaries it appeared that he had outstripped his predecessors in providing a mechanism for the evolutionary process—a mechanism which was neither fanciful nor dependent upon tendencies or laws which defied further investigation.

2.2 The classification of life

Not only the nature of life but the classification of life had been a problem discussed from antiquity. It was a matter of great interest to establish the resemblances and differences between all living things. How could some order be brought to bear on the enormous diversity of life? Was it possible to find criteria whereby all living organisms could be allocated to a particular niche in a classificatory scheme? Aristotle had applied himself to this very problem and concluded that it was indeed possible although there were serious difficulties with ambiguous organisms like the sponge. If complexity of structure and complexity of function were the fundamental criteria then it was possible to arrange all living organisms in one great chain from the lowliest vegetable through the animals to man who was gifted with a rational soul. The apparent vitality of each organism was determined by its own rung on the ladder, and the capacity for motion gradually increased as the scale was ascended. So satisfying was this approach to classification that it has been called 'the greatest synthetic scheme in pre-Darwinian biology'[1] and among Christian commentators it could be extended to embrace even the angels.

Behind all attempts at classification there was a further problem which was brought into focus during the eighteenth century. This concerned the distinction between *natural* and *artificial* methods. The distinction itself was not always drawn in the same way, but broadly speaking a *natural* system of classification was one designed to bring out all the known resemblances, all the observed affinities between living things. The problem with this method was the sheer complexity of the enterprise, the enormous number of available criteria (external form, possession of specific organs, etc.), and the difficulty of establishing a unified scheme on the basis of inductive inference from all known specimens. Those who endeavoured to simplify the problem could look for an *artificial* method based on the selection of a small number of elementary characteristics. This is precisely what the renowned Swedish botanist, Linnaeus, intended when he divided all plant species into twenty-four classes according to their sexual anatomy, to the relationship between their stamens. His system was artificial in the sense that he considered some, not all, of the properties, and in the sense that it placed together plants which, to all appearances, were dissimilar. Thus the French naturalist, Buffon, could complain that it was a defect of Linnaeus' scheme that it confounded the mulberry with the nettle, the strawberry with the rose. There is a sense in which this criticism was wide of the mark, for Linnaeus was well aware of the artificiality of his system.

More telling objections came from Linnaeus himself who discovered not only that stamens could vary among different species of one genus, but also that the flowers of certain plants could exhibit a different number of stamens at different stages of growth. Nor did Linnaeus entirely succeed when he tried to conceive a natural system based upon general appearances—partly, it has been argued, because he could never relinquish his belief that the sexual parts of plants were their most essential features.

Why was he so convinced that this was so? The answer appears to be that, like Erasmus Darwin, Linnaeus subscribed to the botanical analogy, to the belief that there were sets of material analogies between plants and animals: 'To illustrate the generation of plants, we must take our first lights from the animal kingdom and pursue the chain of nature till it leads us to vegetables.'[2]

But what was this 'chain of nature' that could promote the analogical argument? It was in fact nothing less than an elaboration of the linear classification of Aristotle which had become so deeply scored on the western mind that it could determine both natural and artificial systems of classification, as in the case of Linnaeus, and dominate man's conception of his place in nature, as it did for many of Linnaeus' contemporaries. The history of 'the great chain of being' has been brilliantly written by Professor Lovejoy in his book of that title, which you would do well to consult at some point during the course. For the moment you might like to study the chain itself as it was assembled by the French systematist, Charles Bonnet, in 1764 (see overleaf).

Did you observe that it is not just a chain of life, but a chain of natural beings including both inanimate and animate things? Did you notice the fine links in the chain—the way in which truffle could mediate between stones and mushrooms, the polyp between plants and animals, the bats and flying squirrels between the birds and the quadrupeds? Is it surprising that such filigree gradation should evoke a sense of wonder and awe? This last question is not out of place since perhaps the most important thing that must be said about the idea of the great chain is that it became far more than a backbone for biological classification: it was a concept which could be sustained by a set of religious preconceptions about the way in which the Creator had structured his creation. You can see this for yourself if you read the short discussion by Soame Jenyns in *Anthology I*, pp. 297–302.

[1] P. C. Ritterbusch, *Overtures to Biology*, Yale University Press, 1964, p. 58.

[2] Linnaeus, *Sexes of Plants*, 1760, 15; cited by Ritterbusch, *op. cit.*, p. 110.

TABLE 1 *Idée d'une échelle des êtres naturelles**

MAN
monkey
tortoise —— QUADRUPEDS
crocodile flying squirrel
sea lion bat
sea calf ostrich
hippopotamus BIRDS
whales amphibious birds
aquatic birds
? flying fish
FISH
? eels and creeping fish
water serpents
? crab REPTILES
crayfish slugs
SHELLFISH
lizard pond mussel
frog lime-secreting worms
INSECTS
worms
polyp
sensitive plants
trees
shrubs } PLANTS
herbs
lichens
molds
mushrooms and agarics
truffle
stones composed of layers, fibres, and filaments
unorganized stones
CRYSTALLINE SALTS
vitriols
SEMIMETALS [nonmalleable metals]
MALLEABLE METALS
sulphur and bitumens
compound earths [pure earths united with oils, salts, sulphurs, etc.]

PURE EARTH

WATER

AIR

ETHEREAL MATTER

* From Charles Bonnet, *Contemplation de la nature*, *1* (1764), 33–69. Reproduced in Ritterbusch, *op. cit.*, p. 71.

After you have read this discussion, state how you would use this source to the full in order to demonstrate the theological connotations of the 'Great Chain'.

Jenyns' account could be used to indicate the following connotations of the great chain:

1 The finely wrought, almost imperceptible, links in the chain could be seen as evidence of God's *infinite* wisdom and power (see Units 9–10, Section 5).

2 The fact that the chain was continuous, with each successive link superior to the one before, could be associated with the belief that God had created every possible creature in order to diffuse all *possible* happiness.
3 By the combination of instinct, and the reason which is man's superior asset, he can be taught belief in God, in a future state and in the difference between good and evil.
4 The very construction of the chain, with its 'beautiful subordination', gives us the clue as to *how* it was constructed.
5 By analogy, the chain could be extended to include higher beings, extra-terrestrial life and the angels.

Jenyns was writing neither as a committed Christian nor as a typical deist. Indeed he incurred the wrath of Dr Johnson for his facile account of human misery. Manipulating the chain to his advantage Jenyns had argued that between man and the Supreme Being

> there may be numberless intermediate beings who have power to deceive, torment, or destroy us for the ends of only their pleasure or utility, who may be vested with the same privilege over their inferiors, and as much benefited by the use of them, as ourselves.[1]

To which Dr Johnson replied with biting satire:

> many a merry bout have these frolick beings at the vicissitudes of an ague, and good sport it is to see a man tumble with an epilepsy, and revive and tumble again, and all this he knows not why . . .[2]

As long as the chain was used to underwrite the belief that this is already the best of all *possible* worlds, as it was in Jenyns' case, it sponsored a sterile optimism which could easily lead to complacency in the context of social reform. Though there were 'numberless imperfections inherent in all human governments', they were 'imputable only to the inferiority of man's station in the universe, which *necessarily* exposes him to natural and moral evils.' In a more biological context the great chain was bound to affect speculation about the possibility of organic evolution. Study the four extracts below, and then write a short paragraph in answer to this question: how might belief in the chain of being have affected speculation about the evolution of new creatures from more primitive ones?

> each moss,
> Each shell, each crawling insect holds a rank
> Important in the plan of Him, who fram'd
> This scale of beings; holds a rank, which lost
> Wou'd break the chain, and leave behind a gap
> Which nature's self would rue.[3] [STILLINGFLEET]

> From the supposition of a Scale of Beings, gradually descending from perfection to nonentity, and complete in every intermediate rank and degree, we shall soon see the absurdity of such questions as these, Why was not man made more perfect? Why are not his faculties equal to those of angels? Since this is only asking why he was not placed in a different class of beings, when at the same time all other classes are supposed to be full.[4] [LAW]

> The wide interval which separates man from the true quadrupeds is filled by the apes and the animals which most nearly approximate them, of which the species are minutely

[1] B. Willey, *The Eighteenth-Century Background*, Peregrine, 1962, p. 56.
[2] Willey, *op. cit.*
[3] B. Stillingfleet, *Miscellaneous Tracts*, 1759, 128; cited by Ritterbusch, *op. cit.*, p. 69.
[4] Cited by A. O. Lovejoy, *The Great Chain of Being*, Harper, New York, 1960, p. 216.

graduated. . . . We ascend as it were by so many steps to a superior or principal species which so closely resembles man that he has received the name of *orang-outang*, or savage man. It is here above all that it is impossible to fail to recognize the graduated progression of beings; it is here above all that is verified the famous axiom of the German Plato, *Nature makes no leaps*. . . . The contemplator of Nature arrives with surprise at a being resembling man so nearly, that the characters which distinguish them seem less the characters of species than of mere varieties.[1] [BONNET]

In the prodigiously varied sequence of the animals below man, I see Nature in labour advancing fumblingly towards that excellent being who crowns her work. However imperceptible the progress which she makes in one step, that is, in each new production, in each variation upon the original design which she achieves, nevertheless the advance becomes clearly sensible after a certain number of metamorphoses. . . . All the varieties intermediate between the prototype and man I regard as so many essays of Nature, aiming at the most perfect, yet unable to attain it except through this innumerable sequence of sketches. I think we may call the collection of the preliminary studies the apprenticeship of Nature in learning to make a man.[2] [ROBINET]

It is clear from the Stillingfleet passage that belief in the scale of beings could readily endorse the supposition of a fixed creation (see Units 9–10, Section 5). The possibility of extinction is evidently excluded. Nor was there room for new species, or old ones with improved faculties, as the extract from Law confirms: even if nature had the means of producing new species there would be no place for them in the scheme of things. All possible, all viable, classes were already full. But this did not necessarily mean that the respective classes were utterly distinct. In fact the continuity principle encouraged Bonnet to speculate whether man and ape might belong to the same species. This of course implied nothing about a process of 'evolution', though the last extract, from Robinet, shows how the idea of the chain could be adapted to accommodate such a process. The chain could be laid out along a time axis with each successive link an almost imperceptible advance towards man. Some of the earliest ideas on 'evolution' were of this kind. Even Lamarck, though he rejected the idea of a *single* chain—on the ground that he could see no continuity between plants and animals—nevertheless retained the notion of an inherent tendency towards greater complexity which was directed towards man.

2.3 The concept of species

Associated with problems of classification were difficult questions concerning the notion of species. A species could never be defined as a group of *identical* individuals since it was well known that small variations occur among specimens which are obviously of the same stock. But were there criteria to distinguish between one species and another, to distinguish between species and mere varieties of one and the same species? Behind these questions was one still more fundamental. Were there real, distinct species in nature at all, or was the concept of species a purely artificial device to facilitate classification? Those who argued for the reality of species could conceive of ideal, fixed, morphological types from which individuals could deviate in superficial respects. But those who adopted the nominalism of the French naturalist Buffon, argued the artificiality of all attempts at classification, and insisted that a definition of 'species' could never be more than pragmatic. As Buffon himself once remarked, nature knew nothing of families, only individuals.

Figure 2 Buffon, engraving by Meyron from an original by Romney (Mansell Collection).

In what sense did this debate have religious connotations? It did inasmuch as belief in the reality of fixed species appeared to be ratified by the book of Genesis. Aristotle had observed that 'in the normal course of nature the offspring which a male and a female of the species produce is a male or female of that same species.'[3] This was not

[1] Lovejoy, *op. cit.*, p. 235.
[2] Lovejoy, *op. cit.*, p. 280.
[3] Bentley Glass, 'The Germination of the Idea of Biological Species', in Bentley Glass (ed.), *Forerunners of Darwin*, Johns Hopkins, 1959, p. 31.

in itself a definition of 'species', but the truth of the observation encouraged later naturalists to seek a definition in these terms. John Ray in the late seventeenth and Linnaeus in the eighteenth century, examined reproductive capacity as a criterion for the determination of genuine species: only those individuals which could be successfully mated were representatives of the same species. And for many this became more than a pragmatic definition.

Read the first chapter of Genesis, verses 21 to 25, and see if you can say why.

These verses could be interpreted to mean that each of God's creatures could only reproduce its own kind. There was a divine imperative behind multiplication each 'after his kind'.

This may help to explain why Linnaeus should affirm that there are as many species as God originally created. It is true that he later declared species to be the work of time, but his authority was frequently invoked in subsequent generations when the fixity of species required a seal of approval. Nor was the ratification of fixed species derived from the Bible alone. If one believed that God had adapted each creature to an environment that was essentially stable, then it was hard to believe that individual variations were not confined within well defined limits. From their otherwise very different standpoints the comparative anatomist Cuvier, the geologist Lyell, and the mathematician and philosopher William Whewell were agreed on this point. There were severe limits to variation.

It is here that the relevance of the 'species' problem to 'evolutionary' speculation should be clear. If each species had its archetype, or if belief in the fixity of species enjoyed religious sanction, then speculation concerning the large scale *transformation* of species was almost bound to appear radical if not positively rash. Of course a *denial* of the reality of species did not imply any process of progressive transformation, but there is no doubt that it would help to make such a process more credible. It is, however, a measure of the *in*credibility of that process that Buffon himself, who in the opening discourse of his *Histoire Naturelle* had questioned the reality of species, should subsequently adopt a pragmatic definition which as surely excluded marked transformation as did any belief in archetypes. He could see that a hypothesis of common descent might explain the obvious affinities between horse, ass and zebra, but it was a hypothesis he still felt bound to reject. Having defined species in terms of reproduction, having adopted as his criterion the production of *fertile* offspring, Buffon maintained that there was an insurmountable barrier to transformation—a barrier raised by the sterility of hybrids. There could be variation in the history of a species—enough to suggest that elephants might be modified mammoths—but that variation was restricted: 'one can always draw a line of separation between two successions of individuals that can not reproduce with each other. This distinction is the strongest that we have in natural history.'[1]

2.4 The process of reproduction

As you have just seen the criterion of successful reproduction could be used to gain access to the concept of species. Indirectly, theories of reproduction were therefore associated with the plausibility of 'evolutionary' schemes. But as you will see in a moment, the connection was sometimes more direct.

Of all biological problems that of reproduction has been the most recalcitrant. How are parental characteristics transmitted to the next generation? It is only in recent years that biochemists have cracked the code of inheritance with their models of RNA and DNA. In previous centuries the development of an embryo into a likeness

[1] P. L. Farber, 'Buffon and the Concept of Species', Journal of the History of Biology, 1972, 5, 2, 267.

of the adult was so remarkable a process that the prospect of finding a mechanism was daunting indeed and even the relative importance of the roles of the two parents was debated. In the late seventeenth and early eighteenth centuries there were, broadly speaking, two approaches to the problem: the one, dating from Aristotle and reaffirmed by Harvey, is usually given the name of 'epigenesis'; the other that of 'preformation'. Those who subscribed to epigenesis insisted that the embryo developed from a rudimentary mixture of parental semens by a process of differentiation, increasingly complex structures developing with time. Those who adopted the principle of preformation insisted that the embryo was already in some measure preformed within a parent organism: the embryo 'enlarged' rather than 'evolved'. The preformationist position might seem rather crude, but it had a particular appeal for some though by no means all who sought to explain physiological processes in mechanistic terms. The fact is that it was virtually impossible to find mechanical analogues for an apparently teleological process which involved increasing complexity and differentiation. If directive vital forces were prohibited then a pre-packed offspring was one possible answer!

The debate cannot be reduced, however, to the confrontation of 'vitalism' and 'mechanism'. Experimental observations and their interpretation were also involved. If one accepted the preformation principle there was the additional problem of deciding whether the offspring were preformed in the male sperm or the female egg. There had been some resistance to the first of these options, since it implied a great waste of life, but during the eighteenth century it was the latter option which prevailed, especially after the investigations of Von Haller, Bonnet and Spallanzani. Haller appears to have been converted to the 'ovist' theory as a consequence of observations which were eventually published in 1758. While studying the development of embryos he was struck by the fact that each organ was already functioning the moment it became visible. The heart would beat before any blood could be seen passing through it. Since the operations of the bodily organs were known to be mutually dependent, it was reasonable to suppose that all the essential organs were present from the start—especially when the use of preservatives disclosed the outline of otherwise invisible parts.[1] As with the doctrine of spontaneous generation, the doctrine of preformation was hard to disprove. It was challenged by the German biologist K. F. Wolff who, in a thesis dedicated to Haller, claimed to have observed the formation of the organs of a chick from certain 'vesicles' which were themselves produced from the original formless jelly. But did Haller capitulate? In fact he was prepared to accept the observations but reaffirmed that in the primitive jelly the organs were simply too minute to be seen. Even when Spallanzani produced a detailed description of spermatozoa, it was possible to argue, as he himself did, that they were small parasites and in no way essential to conception!

We shall not pursue the religious connotations of this debate, though it is a useful reminder of the non-equivalence of religious and vitalistic beliefs that Bonnet could develop a theory of preformation which was mechanistic in intent and yet grounded in God's creative power. From 1764 until his death, thirty years later, Bonnet argued that only God could create living things; that the germs of all plants and animals had been introduced at the Creation; but that, since then, all natural processes including the successive expansion of germs had been governed by the forces of physics and chemistry.[2]

To ask whether the preformation/epigenesis debates had any bearing on evolutionary speculation is, however, a far more important question, and one which you should now try to answer for yourself. Can you say which of the two presuppositions concerning the process of reproduction was likely to be less inhibitive to speculations concerning the development of one species from another? How would you justify your answer?

A belief in epigenesis would be less inhibitive, since acceptance of preformation would

[1] E. Gasking, *The Rise of Experimental Biology*, Random House, 1970, p. 73.
[2] Gasking, *op. cit.*, p. 75.

tend to reinforce a belief in the fixity of species. If each generation was in some sense 'contained' in the germs of the one preceding, then there could scarcely be scope for variation. Pre-packed antelopes would always remain antelopes: they could not develop into giraffes.

You might perhaps have justified your answer in a second way. Perhaps an epigenetic account of a developing foetus could serve as an analogy for the progressive development and differentiation of living forms? The word 'evolution' was commonly and primarily used to refer to the development of an embryo. Could it be applied to the history of life on earth? Though the analogy looks tenuous to us it did eventually find its way into the 'evolutionary' scheme of Chambers who, in his *Vestiges* (1844) had convinced himself that 'each animal passes, in the course of its germinal history, through a series of changes resembling the *permanent forms* of the various order of animals inferior to it in the scale'. Nor was man to be exempt: 'His organization gradually passes through conditions generally resembling a fish, a reptile, a bird, and the lower mammalia, before it attains its specific maturity.'

A different alliance between 'evolutionary' speculation and the rejection of preformation had found expression exactly a hundred years before the publication of Chambers' book. The physicist and philosopher, Maupertuis (1698–1759), who had done much to popularize Newtonian ideas in France, launched an attack on preformation which is said to have been inspired by the contemplation of an albino Negro boy displayed before Parisian society in 1744. What could be the source of so remarkable a variation? Against those preformationists who had explained individual variation with reference to accidental factors, such as an abnormal food supply during development, Maupertuis protested that such a theory was powerless to explain why it was that particular variations could be transmitted from a parent to an offspring or indeed recur in a later generation. His own explanation was that both sexes produced a semen which contained particles drawn from every component of the body—an idea that had been first proposed by the Hippocratic group in ancient Greece. After intercourse there was an attraction between similar particles which would come together in the uterus and so construct the organs of the offspring. The phenomenon of the white Negro could then be ascribed either to a spontaneous variation in the parental particles or to the reactivation of particles derived from a white ancestor. Having once allowed the possibility of spontaneous variation, Maupertuis indulged his imagination to the full:

> Could one not explain by that means how from two individuals alone the multiplication of the most dissimilar species could have followed? They could have owed their first origination only to certain fortuitous productions, in which the elementary particles failed to retain the order they possessed in the father and mother animals; each degree of error would have produced a new species and by reason of repeated deviation would have arrived at the infinite diversity of animals that we see today; which will perhaps still increase with time, but to which perhaps the passage of centuries will bring only imperceptible increases.[1]

It is speculations of this kind which have tempted historians to scan the literature for 'anticipators' of Darwin, for 'precursors' who, wittingly or unwittingly, grasped the rudiments of Darwin's theory well in advance of the master himself. Gross exaggeration and gross distortion have been the consequences, as you will see in Section 2 below. Ingenious though Maupertuis' deviation theory undoubtedly was, it suffered from defects which were all too clear to his contemporaries. Was it really possible for an attractive force, analogous to gravitation, to assemble and co-ordinate the seminal particles into organs of such intricacy? Maupertuis had to say that his force was possessed of a memory which enabled it to remodel the parental structures, but such a twist was hardly consistent with a mechanist programme! This should not mean, however, that Maupertuis' speculations are of no interest to the historian. Though it is not clear that they led anywhere, they serve as a useful reminder that any cogent

[1] Gasking, *op. cit.*, p. 64.

theory of the transformation of species would have to give a plausible account of both the source and the transmission of variations. As you will discover in Unit 13, even Darwin himself was to have problems in these areas.

2.5 The testimony of the rocks

In any discussion of the background to evolutionary speculation it would be almost impossible to overestimate the significance of the fossil record. During the eighteenth century, knowledge of the fossil record was sparse, and for perfectly sound reasons the history of life was believed to be a matter of a few thousand years. Once it became clear, as it did to Cuvier, that the science of palaeontology could reveal the pattern of prehistoric life, the characteristic flora and fauna of distant epochs, a profound consciousness of the long corridor of time began to affect explanations of the pattern itself. The discoveries of the geologists also raised several questions of interpretation which were frequently discussed in the context of their theological implications. This is ground that you have already covered in Unit 11 and in television programmes 7 and 8, but it might be useful as a revision exercise if you could state below what these controversial questions were.

Figure 3 Cuvier and fossil remains, from the painting by Chartran (Mansell Collection).

What were the geological issues which were seen by contemporaries to have religious implications?

My list would include the following:

1 The evidence for the extinction of many species.
2 The evidence for the enormous age of the earth.
3 The problems concerning whether fire or water was the primary geological agent, and the special case of Noah's flood.
4 The interpretation of the breaks in the fossil record, and the concomitant question whether the history of the earth should be written as a series of distinct epochs separated by major terrestrial upheavals.
5 The evidence for a sense of direction in the fossil record—a direction towards creatures of ever greater complexity, and culminating in man.
6 The general question whether it was possible to give a complete account of the history of the earth without having to postulate any interruption to the continuous action of the natural causes still prevailing.
7 The problems concerned with the exegesis of Genesis: did the geological evidence support or contradict the *sequence* of creative acts recorded in the Bible?

Each of these issues presented problems for certain kinds of religious belief, but as Professor Hooykaas and Dr Rudwick have suggested, there are no sharp lines to be drawn between revolutionary geologists and clerical amateurs. The evidence for extinction presented a challenge to those eighteenth-century authors who had argued that God had brought into co-existence every *possible*, every *viable* creature; but it could not destroy the plenitude argument altogether. Given that the earth's surface had provided an ever changing environment, God might still have pre-adapted a maximum number of species to enjoy each successive, though transient, environment. Nor was it the case that every 'evolutionist' made a meal out of extinct species: the scheme of Lamarck, for example, took no notice of extinction. The increased age of the earth certainly offended those who had set their chronometers by the standard of Ussher, who, from Biblical data, had fixed on 4004 B.C. as the year of Adam's 'birth', but there is little evidence to suggest that practising geologists were, in this respect, ever seriously inhibited. It was always possible to argue that the 'days' of Genesis referred to vast periods of time, or with Buckland, to suggest that there was an immense interval of time between the original creation and the first Genesis day. When fire displaced water as Earth's most powerful sculptor there were protests from some who had deduced their geology from the first page of their Bibles, but it would be naive to suppose that Neptune and the devout fought Vulcan and the devil. Hutton may have insisted that the rocks themselves revealed no vestige of a beginning, no prospect of an end, but that was not the same as asserting that there had been no beginning and would be no end. In fact, behind Hutton's theory of the earth was a sense of design and order no less pronounced than among those who may have received him with suspicion (see Unit 11).

Similar complications must be taken into account in any worthwhile discussion of the so-called catastrophist/uniformitarian debate. You will hardly need to be reminded that the 'bad press' given to the 'catastrophists' by some historians is no longer credible. The fact that gigantic boulders and disjointed strata testified to violent upheavals cannot mean that those who advocated discrete epochs separated by 'catastrophes' did so merely to save the notion of miracle. Since God could be assumed to work though 'natural' causes, the concept of 'catastrophe' did not *necessarily* imply intervention at all. Nor was Buckland's provisional identification of the last catastrophe with Noah's flood typical of the best 'catastrophist' geology. On the question of *direction* in the fossil record religious interests were certainly involved, but once again in ways which would defeat any *a priori* generalization. One might expect that a sense of direction in the history of life would point to the possibility of progressive evolution, with consequent embarrassment to an orthodox belief in the uniqueness of man. Certainly there is evidence that Charles Lyell saw things in this way, and this may help to explain his opposition to 'directionalism'. On the other hand a pattern of increasing complexity culminating in man could be used to argue for a Providential blueprint: direction surely presupposed a Director? Sensitive to this line of argument the French materialists who followed in the footsteps of Diderot and d'Holbach (see Unit 7) had been inclined to *reject* any 'evolutionary' theory that implied necessary improvement rather than degeneration. It should not therefore come as a surprise to learn that Lamarck's theory of 'evolution' was deistic rather than atheistic. And, as you have already seen in Units 9–10 and 11, the belief in a direction to the history of life could be married with belief in progressive creation, as it was in the outlook of Buckland, Sedgwick and Miller.

Concealed within the pages of *Gillispie* (set book) is an error which stands as a warning against the oversimplification of the story. Read first of all what Gillispie has to say on page 104 with reference to a 'whole generation of scientific theologians'. Pay particular attention to the content of footnote 23 (p. 279). Then study Gillispie's characterization of Lyell's uniformitarianism on page 135.

What comments would you make if you were now told that the author of the article in *Quarterly Review*, vol. xxxiv (1826), 507–540, was none other than Lyell himself?

Your comment should be something like this; Gillispie sets up an antithesis between the 'uninspired' theological science of Buckland which, compared with the 'optimistic materialism' of Lyell's uniformitarianism, was 'hopelessly . . . out of key with the times'. Unfortunately the very article to which he refers in order to typify the natural theology of Buckland and his school was written by Lyell himself.

When Gillispie wrote *Genesis and Geology*, it was not widely known that Lyell was the author, so perhaps he may be forgiven. But the error of interpretation remains, for Lyell continued to agree with Buckland that the position of coal in stratified rocks was just one example 'of the adaptation of particular means and forces to the accomplishment of certain ends for which the habitable earth has been framed'.[1] In other words Lyell's conception of an earth history that could be written without reference to the action of causes any different in kind or intensity from those we see at work today, was as much indebted as inimical to the natural theology of Paley whom he frequently quoted with approval.

[1] Lyell, *Proceedings of the Geological Society of London*, 1837, **2**, 479–523; cited by M. Bartholomew, 'Lyell and Evolution: An Account of Lyell's Response to the Prospect of an Evolutionary Ancestry for Man', *British Journal for the History of Science*, 1973, **6**, 267.

We have included this critique of Gillispie not to score a cheap point, but to emphasize how warily the historian must tread, and nowhere more so than where the testimony of the rocks is concerned. With the privilege of hindsight T. H. Huxley could claim that consistent uniformitarianism in the history of the earth—in the history of inorganic processes—demanded consistent uniformitarianism in the history of life—in the history of organic processes. But, as Professor Hooykaas has insisted, without that privilege, the parallel could not be so simply drawn. The word uniformitarianism is notoriously ambiguous, and neither Huxley nor Lyell defined precisely what they meant by it. If by 'uniformitarianism' Huxley meant the explanation of the past in terms of causes and processes in action today, then since no one had *observed* the transformation of species today, there could be no sanction for pleading transformation in the past. Not that special acts of Creation could be observed either! If by 'uniformitarianism' Huxley meant the explanation of the past without reference to divine activity, then one might sympathize with his point of view, but to most of his predecessors there were such palpable differences between rivers and rocks on the one hand and man and beast on the other, that his analogy would have looked contrived. And if by uniformitarianism Huxley meant a *uniform* past—a past with no progressive change—then his analogy would have to preclude the progressive evolution of life. It was of course possible to argue for the uniform, continuous *development* of life—for uniformity in the laws of 'evolution'—as Erasmus Darwin had done (see Unit 11). Yet the very fact that Lyell—the foremost geologist of his age—could insist for more than thirty years that the fossil record, properly understood, did *not* reveal any directional change during the history of life, should stand as a warning against the common assertion that the rocks bore testimony to 'evolution'. That testimony came from *some* of their interpreters, but not from the rocks themselves! Lyell may have been out on a limb when he denied progression and direction in the fossil record but such direction as there was did not suggest a pattern of 'evolution'. Charles Darwin himself was to concede that the *absence* of intermediate forms was the gravest objection to his theory.

There was of course, a fundamental question raised by the rocks: how could one explain the *differences* between fossil and living forms? Addressing himself to this question at the turn of the century, Cuvier had identified at least three possible explanations. Can you work out what they must have been?

As Cuvier identified them, the options were these: 'It is a question of finding out whether the species that existed (in the earth's distant past) have been entirely destroyed, or if they have only been modified in their form, or if they have simply been transported from one climate to another.'

These were the options, only the second of which *demanded* a mechanism for the modification of species. Nor was this the choice of Cuvier himself, who having first relied on global catastrophes to explain extinction on a grand scale, subsequently qualified his thesis by appealing to migration for the repopulation of particular areas following catastrophes more local in their effects. Lamarck, on the other hand, was far more willing to espouse that crucial second option—in conjunction with the third. This would seem to suggest that one's choice of option was determined by considerations that were not exclusively geological.

In Lamarck's case an aversion to 'catastrophes', a belief in the order and harmony of nature which made it difficult for him to accept extinction at all, and his expertise in conchology (the study of shells) steered him towards the possibility of species transformation.

After all, in the case of marine organisms it was virtually impossible to demonstrate extinction. Can you see why this was?

Obviously one could argue, as had the earlier French naturalist Bruguière, that the living counterparts of some nautical fossils were simply inaccessible, fathoms deep in the ocean!

As a consequence of his own research Lamarck was able to locate living species which, in more than forty instances, did resemble fossil shells—not exactly, but sufficiently to place them in the same genus. From here, according to the author of one recent study, it was but a short step to the doctrine of transformation:

> Unwilling to admit that some major catastrophe might have caused extinction, unable to conceive of a natural mechanism by which extinction might take place, but granted that there were differences between fossil and living forms that probably would never be explained simply by the discovery of living analogs of the fossils, Lamarck was quite naturally led to a belief in species mutability.[1]

Such a belief, once entertained, could of course be endorsed by further geological considerations. If species were adapted to their environments, and if their environments had been subject to gradual changes, could the consequent pressures, exerted on the species, induce responses which might lead to a gradual transformation? In framing answers to this question Lamarck propounded a theory of evolution which could be made to sound quite plausible. As he declared in 1809:

> Now, the proper sequence of things to consider in all this is as follows:
> 1 Every fairly considerable and continuing change in the circumstances of any race of animals brings about a real change in their needs (*besoins*).
> 2 Every change in these needs necessitates different actions to satisfy the new needs and, consequently, different habits.
> 3 Every new need, necessitating new actions to satisfy it, requires of the animal that it either (a) use certain parts more frequently than it did before, thereby considerably developing and enlarging them, or (b) use new parts, which their (new) needs have imperceptibly developed in them, by virtue of the operations of their own inner sense (*par des efforts de son sentiment intérieur*).[2]

Here was the basis of a theory of transformation which, though repeatedly discredited in the decades that followed, haunted the imagination of all who sought to construct a history and classification of life.

2.6 The uniqueness of man

'I remember that it was the conclusion he came to about man that fortified me thirty years ago against the great impression which his arguments at first made on my mind.' Thus Charles Lyell wrote to Charles Darwin, 15 March 1863, some thirty-six years after first reading Lamarck. However persuasive, or however easily it could be discredited, Lamarck's theory was rarely judged on its scientific merits alone. If species evolved by progressive transformation, what were the implications for man himself? Of Lamarck's argument Lyell had this to say: 'if pushed as far as it must go, if worth anything, it would prove that men may have come from the Orang-Outang.'[3]

If pushed as far as it must go: here was a criterion by which all theories of evolution were almost bound to be judged. What they had to say of the origin of species would seem inconsequential compared with what they refrained from saying about man himself:

[1] R. W. Burkhardt, Jr., 'The Inspiration of Lamarck's Belief in Evolution', *Journal of the History of Biology*, 1972, **5**, 2, 424.
[2] H. L. McKinney (ed.), *Lamarck to Darwin*, Coronado Press, Kansas, 1971, pp. 14–15.
[3] Letter of Lyell to Mantell, 2 March 1827; cited by Bartholomew, *op. cit.*, p. 273.

> You ask whether I shall discuss 'man'. I think I shall avoid the whole subject, as so surrounded with prejudices; though I fully admit that it is the highest and most interesting problem for the naturalist.

This was Darwin's strategy as he outlined it to Wallace, and indeed all that he did say of man in the *Origin* was to be condensed into one portentous sentence: 'Much light will be thrown on the origin of man and his history.' The problem was, of course, that what Darwin and the evolutionists took to be light, their opponents took to be a dangerous darkness. To identify all the 'prejudices' with which Darwin felt he had to contend, would be a forlorn venture, but the following reaction of the geologist Adam Sedgwick may help you to understand the problem:

> I have read your book with more pain than pleasure.... There is a moral or metaphysical part of nature as well as a physical. A man who denies this is deep in the mire of folly. Were it possible (which, thank God, it is not) to break [the link between the moral and the material], humanity, in my mind, would suffer a damage that might brutalize it, and sink the human race into a lower grade of degradation than any into which it has fallen.

Compare this response of Sedgwick with his earlier castigation of Chambers (see Gillispie (set book), p. 165). Then state what you consider to be Sedgwick's primary motive for denouncing the transmutation of species.

His primary motive appears to have been a conviction that human *morality* would be jeopardized if 'evolutionary' theories were accepted.

This is an important point because *moral* objections to 'evolution' need not *necessarily* be derived from religious dogma. In Sedgwick's case they obviously were: 'If [Chambers'] book be true ... religion is a lie.' But one could be worried by a threat to human dignity without subscribing to religious dogmas. Thus Lyell could inform Mantell that on reading Lamarck he had felt 'none of the *odium theologicum* which some modern writers in this country have visited him with'.

But he clearly felt *something* which offended his deepest instincts. Man was a moral being, gifted with 'improvable reason'; it was scarcely conceivable that he could have developed from orang, ape or gorilla.

The fate of evolutionary speculation was therefore closely associated with the fate of man himself. There was a widespread fear among moralists of different persuasions that if men were told they shared an ancestor with the apes they would probably stop behaving as men. From our vantage point this may well seem a curious prejudice, but we should have an incomplete picture of the background to Darwin if we failed to recognize that for a Hugh Miller and an Adam Sedgwick the prospect of man's evolution was one with the prospect of moral anarchy. And such a prospect would have obvious political repercussions. The issues were further complicated by the fact that though belief in the uniqueness of man was derived from religious, moral, political and introspective sources, it could for a while be squared with the 'scientific' evidence concerning man's antiquity. When Lyell wrote his *Principles of Geology*, there were no vestiges of humanity strewn among the strata to suggest anything other than a *recent* origin of man. Nor were there 'missing links' to corroborate the speculations of Lamarck. Though Lyell was soon to be criticized for suggesting that mammals had existed from the earliest times, he was not criticized for making an exception of man, whose recent appearance and above all his faculty of improvable reason implied special creation.

In this section we have examined some of the more important conceptual problems which were associated both with religious belief and evolutionary speculation in the period before Darwin. There were of course many others, including the continuing debates concerning 'spontaneous generation'. Recent studies have shown what an ambiguous term this is: it could refer to 'abiogenesis'—the generation of living organisms from inorganic matter; or to 'heterogenesis'—the generation of living organisms from *organic* matter.[1] Whereas the former possibility was largely rejected by the second half of the eighteenth century, the latter was still debated until well into the nineteenth century—thanks in part to that most despised of creatures, the intestinal worm, whose presence could be explained as well in terms of generation as of germination. As you have seen in Unit 7, a belief in spontaneous generation, abiogenesis, could be an alternative to both 'creation' and 'evolution' as an account of the origin of living things (cf. Diderot).

However, among those who did propound a thoroughly evolutionary theory before Darwin a belief in some kind of 'spontaneous generation' was still a desideratum. Can you see why?

A theory of evolution which presumed to explain the development of life as we know it, without reference to *any* supernatural acts of creation, would have to be able to explain the origin of the most primitive organisms from which all other life gradually 'evolved'.

Lamarck eventually came to believe that heat and electricity acting upon inanimate matter would perhaps produce the most primitive organisms, the 'mere sketches of animality' as he called them. And such a belief was part and parcel—perhaps even a precondition—of his theory of evolution, as it was to be for Robert Chambers forty years later. Darwin was to side-step the problem of the origin of the first primordial life, but in so doing he incurred the dismay of the materialists.

You will learn more of the background to Darwin as you proceed with this unit and the next two. In Britain, the political controversies ignited by Malthus' *Essay on Population* have to be taken into account; and, as Dr Richardson argues in Unit 14, a revolution in man's understanding of history was fundamental to his understanding of nature. For the moment, however, you should have a sufficient grasp of the concepts to appreciate the historiographical problems which are the subject of our next section.

Perhaps you would care to check that you could enumerate some of the beliefs prevalent during the eighteenth and early nineteenth centuries which would help to explain why speculations concerning the transformation of species were commonly regarded as mere speculations. Which concepts would you select?

Your list should include the belief that nature was an essentially static machine; that there was a great chain of being which allowed no room for the emergence of new species, or the extinction of old ones; that species were real, distinct, and able to reproduce only after their kind; that offspring were essentially preformed; that the fossil record testified to progressive *creation*; and that man, as a moral being, was unique.

---

3 Some Historiographical Considerations

3.1 Precursors or predecessors?

You may have wondered why the title of this unit bears a question mark. Now is the time to pause and to think about it.

[1] J. Farley, 'The Spontaneous Generation Controversy (1700–1860)', *Journal of the History of Biology*, 1972, **5**, 1, 96.

What danger do you see in interpreting pre-Darwinian biology in terms of Darwin's *precursors*? Refer back to the discussion of whiggish history in Unit 1 if you are in any doubt.

The most serious danger is that of *distortion* if we write history backwards from the vantage point of Darwin's theory. To search the literature for 'anticipations' of Darwin's own ideas could entirely corrupt our understanding of his predecessors, their problems, their methods, and their solutions. To evaluate their work as if they were consciously contributing towards a theory of evolution that was not yet formulated could be profoundly misleading—especially if one assumed at the outset that all 'evolutionists' were enlightened and their opponents dull.

Your answer should have been along those lines. The dangers are, however, even greater than such an answer might suggest. In the first place there is a trap concealed within the word 'evolution' itself: because a predecessor of Darwin chances to use the word, it does not mean that he has found a mechanism for the transformation of species, or even that he believes in transformation at all. As we have stressed before, the most common context in which the word was used in the pre-Darwinian period was probably that of embryology: the evolution of the embryo from egg to offspring.

In the second place, even though a predecessor of Darwin may have articulated ideas which subsequently found their way into the Darwinian synthesis, one must always ask: how important, how significant were these ideas for the man who stated them? Were they integral to a system of his own, or were they merely footnotes, idle speculations, alternative points of view? You will see shortly that Buffon has been hailed as a precursor of Darwin because of his belief that a species could change with time. But the truth is, as Robert Wohl insisted, 'Buffon never took seriously the question of transformism because he considered change in species a special case rather than a basic characteristic.[1] Similar considerations apply to the concept of struggle in nature which can be found in the writings of biologists and philosophers from antiquity. That they may have made passing reference to it does not mean that they had understood the significance of competition as Darwin was later to do.

A third complication is this: though a predecessor of Darwin may have used (as distinct from merely stated or guessed) ideas which Darwin was to make his own, such ideas were not infrequently associated with others which directly contradicted what were to be the main tenets of Darwin's theory. Thus historians have been tempted to portray Lyell as a *precursor* of Darwin for the very good reason that both Darwin and Wallace were deeply indebted to his perspective on the way in which small changes could gradually accumulate during the aeons of time. But since Lyell so steadfastly opposed not only the concept of transformation but also the concept of a progressive pattern in the history of life, the label is hardly conducive to a proper understanding of Lyell. And in the context of 'proper understanding' it is obviously important for the historian to try to discover what the intentions of a scientist or philosopher really were, rather than suppose they were the same as Darwin's, achieved less well. As Jonathan Hodge has recently stressed, the intentions of Robert Chambers were not those of Darwin, whose energies were directed more towards the origin of species and less to the origin of life, more towards the geographical loci of speciation and less to systematic analogies between the 'evolution' of life and the evolution of the embryo.[2]

Finally there is the crucial question of influence. No matter how closely a predecessor might resemble a precursor, if his ideas converted no one, or worse still remained unknown, he could hardly bear comparison with Darwin or Wallace, who developed their arguments with such cogency that those who understood them found it difficult to resist their force. One could hardly imagine a critical biologist accepting the mermaids and mermen of de Maillet (see Unit 7), and for all its publicity Chambers' *Vestiges* contained too many scientific infelicities to sway an informed audience. And how

[1] R. Wohl, 'Buffon and his Project for a New Science', *Isis*, 1960, **51**, 194

[2] J. Hodge, 'The Universal Gestation of Nature: Chambers' *Vestiges* and *Explanations*', *Journal of the History of Biology*, 1972, **5**, 1, 127–51.

much less could those minor advocates of 'natural selection', William Wells, Patrick Matthew and Edward Blyth, have converted their readers to an evolutionary theory when, as McKinney has remarked, they had no such object in mind![1]

These words of warning may seem rather laboured, but you will see their significance as you attempt the questions which follow—the first devoted to the interpretation of Buffon, the second to Cuvier.

3.2 Buffon and transformation

You have met Buffon already. In Unit 7 Dr Goodman discussed some of the problems surrounding the interpretation of his scientific and religious beliefs. Before you proceed further you might care to refresh your memory by re-reading his exposition (6.2.5 and 6.2.6).

In what respect is Buffon said not to be a 'forerunner' of Darwin?

In this respect: he did not appear to believe that higher animal forms, such as the elephant, had evolved by the gradual transformation of more primitive species.

Not all historians have been quite so circumspect, however, and as you will see shortly, exaggerated claims have been the consequence. In the question which follows, you will be asked to write a brief criticism of one such account. In order to prepare yourself you should study Professor Wilkie's interpretation of Buffon which you will find in *Anthology II*, pp. 238–81 (especially pp. 245–59). After you have digested his argument, you should examine the very different analysis[2] reproduced below and then answer the question which follows.

> Linnaeus, I have said, had one great rival in the public affection; this was the Comte de Buffon (1707–88). In 1749 he published the first volume of his huge *Histoire Naturelle*, a set of studies of the living world destined to have a wide circulation and to be translated into many languages. It was gracefully, even entrancingly written, and here and there the author managed, not too conspicuously, to touch upon a number of forbidden topics.
>
> The book was written to appeal to two sorts of readers: those interested in the simple description of animals and those intellectuals who might wish to think about what they saw. We need not expect complete candor on the part of a man writing a century before Darwin. Buffon had doubts, hesitations, and fears. He wrote at times cryptically and ironically. He brought forward an impressive array of facts suggesting evolutionary changes and then arbitrarily denied what he had just been at such pains to propose. It is not always possible to determine when he was exercising an honest doubt of his own and when he was playing a game. In any case he could not leave this dangerous subject alone. It fascinated him as, a century later, it was to fascinate Darwin. He had devised a theory of 'degeneration'. The word sounds odd and a trifle morbid today, because we are in the habit of thinking of life as 'evolving', 'progressing' from one thing to another. Nevertheless, Buffon's 'degeneration' is nothing more than a rough sketch of evolution. He implied by this term simply change, a falling away from some earlier type of animal into a new mold. Curiously enough, as his work proceeded, Buffon managed, albeit in a somewhat scattered fashion, at least to mention *every significant ingredient which was to be incorporated into Darwin's great synthesis of 1859*. He did not, however, quite manage to put these factors together. Specifically they may be analyzed as follows:
>
> 1. Buffon observed a tendency for life to multiply faster than its food supply and thus to promote a struggle for existence on the part of living things. 'Nature', he said, 'turns

[1] McKinney, *op. cit.*, p. 7.
[2] L. Eiseley, *Darwin's Century*, Gollancz, London, 1959, pp. 40–45.

upon two steady pivots, unlimited fecundity which she has given to all species; and those innumerable causes of destruction which reduce the product of this fecundity. . . .'

2. He recognized that within a single species there were variations in form. In domestic plants and animals these variations were often heritable, so that by careful selection the stock could be improved and the direction of the improvement controlled. 'There is,' he wrote, 'a strange variety in the appearance of individuals, and at the same time a constant resemblance in the whole species.' He recognized 'our peaches, our apricots, our pears' to be 'new productions with ancient names. . . . It was only by sowing and rearing an infinite number of vegetables of the same species, that some individuals were recognized to bear better and more succulent fruit than others. . . .' Similarly he noted that in the case of the domestic hen and pigeon 'a great number of races have been lately produced, all of which propagate their kinds.' 'In order to improve Nature,' he commented in another volume, 'we must advance by gradual steps.'

3. Buffon was impressed by the underlying similarity of structure among quite different animals, an observation which is a necessary prelude to tracing out ancestral relationships in the fossil past. 'There exists,' he said, 'a primitive and general design, which may be traced to a great distance, and whose degradations are still slower than those of figure or other external relations. . . .'

He philosophized warily that among the numerous families brought into existence by the Almighty 'there are lesser families *conceived by Nature and produced by Time*.' Such remarks, woven into the web of orthodoxy, at times grow bolder, as when he suggested 'that each family, as well in animals as in vegetables, comes from the same origin, and even that all animals are come from one species, which, in the succession of time, by improving and degenerating, has produced all the races of animals which now exist.' 'Improvement and degeneration,' he had earlier remarked, 'are the same thing; for they both imply an alteration of original constitution.' Though Buffon was quick to add something to satisfy the ecclesiastical authorities after such a remark, it is interesting to observe that in repeating all creatures have really been specially created, he says '*we ought to believe* that they were then *nearly* such as they appear at present.' It is obviously a most grudging concession.

4. Buffon foreshadowed in some degree the uniformitarianism of James Hutton at the end of the century. Like De Maillet he sought natural explanations for the formation of the earth and for geological events. After having listed the innumerable effects of rain, rivers, winds, and frost he remarked perceptively, 'We do not pay any consideration that, though the time of our existence is very limited, nature proceeds in her regular course. We would condense into our momentary existence the transactions of ages past and to come, without reflecting that this instant of time, nay, even human life itself, is only a single fact in the history of the acts of the Almighty.'

Buffon anticipated the need of a greatly lengthened time scale in order to account for the stratification of the planet and the history of life upon it. 'Nature's great workman,' he said, 'is time.' By modern standards, of course, his estimates of the antiquity of the globe are very constricted but in his own time they were unorthodox. He thought that it had taken some seventy-two thousand years for the globe to cool from an incandescent state sufficiently to allow for the appearance of life. He assumed that the heat of the globe was imperceptibly diminishing. Further, he calculated that roughly another seventy thousand years would elapse before the planet was so chilled as to be unable to sustain life on its surface.

5. He accepted the fact that some of the animal life of the earth had become extinct. This he ascribed to the cooling of the earth which had eliminated the warmth-loving fauna of an earlier day. He thought that many existing species would, in time, perish for the same reasons. He recognized the bones of mammoth as being those of extinct elephants and foresaw the value of paleontology. 'To know all the petrifactions of which there are no living representatives,' he remarks, 'would require long study and an exact comparison of the various species of petrified bodies, which have been found in the bowels of the earth. This science is still in its infancy.' By this means, however, man, through the use of comparative anatomy, might be enabled to 'remount the different ages of nature.' It will eventually be possible, Buffon thought, to place milestones 'on the eternal route of time.'

6. Buffon also recognized the value of an experimental approach to evolutionary problems. The relations between species, he contended, could never be unraveled without long continued and difficult breeding experiments. 'At what distance from man,' he hinted slyly, 'shall we place the large apes, who resemble him so perfectly in conformation of body . . . ? Have not the feeble species been destroyed by the stronger, or by the tyranny of man . . . ?' Although in some passages he was careful to maintain the distinctive qualities of man, that his 'animal body' had been infused with a divine spirit, he also remarked at a convenient point: 'You unjustly compare, it may be said, an ape who is a native of the forests

with the man who resides in polished society. To form a proper judgement between them, a savage man and an ape should be viewed together; for we have no just idea of man in a pure state of nature.' Buffon then gives a revolting picture of the savage Hottentot so frequently described by the eighteenth-century voyagers and ends by commenting: 'There is as great a distance between man in a pure state of nature and a Hottentot, as there is between a Hottentot and us.'

There can be noted in these passages a tendency which will be seen to descend into the Darwinian era and to find at least faint expression in Darwin's own work. I refer to the preference for continuing the use of the morphology of the living as a key to descent rather than to wait upon paleontology. This inclination was enhanced because of the long use of the Scale of Being, which, of course, had always involved only living forms in a permanently linked relationship. Furthermore, since the fossil past was still little known and its length inadequately perceived, emphasis naturally continued to lie upon existing animals and their relationships. Thus even the pioneer evolutionists tended to see closer phylogenetic relationships in the present than actually existed between European man, Hottentot, and orang. In a few passages Buffon seems to be struggling to free himself from the living comparative ladder, from living 'ancestral forms.' At moments he glimpses the value of family trees and collateral lines of descent. Nevertheless, though he had a premonition that paleontology would prove valuable, he was not in a position to realize its scope or extent; thus his evolutionism was essentially trapped within the present.

7. Buffon was one of the first biologists to sense the significance of animal and plant distributions. He observed marked differences between the faunas of the New and Old World tropics. He also perceived that the northern, Holarctic region was more nearly similar in its fauna and most similar when Asia and North America adjoined each other. As we have remarked earlier there were seventeenth-century writers who had puzzled over these differences between the fauna of the New and Old Worlds. Buffon, however, answered the question in a truly modern fashion. He said of the New World species which differed from those of the Old Continent: 'They . . . have remote relations, which seem to indicate something common in their formation, and lead us to causes of degeneration (i.e., evolution) more ancient, perhaps, than all others.' Thus Buffon had glimpsed that animals, instead of diffusing from the ark on Ararat, had, partially at least, originated in the areas where they were now to be found. They had, in other words, arisen by modification from ancestral forms previously inhabiting the same region. Here we can observe the first premonitory formulation of the Law of Succession which was to be demonstrated paleontologically by Clift, Owen, Darwin, and others, later on in the nineteenth century.

8. Last, we may note that Buffon also had distinguished, though briefly and uncertainly, something of that world of eternal imperfection and change which, later on, was to fascinate Darwin. He had seen 'doubtful species,' 'irregular productions,' 'anomalous existences.' He had stared into the magic mirror of nature and, like the gods, had seen for a moment the cloud forms streaming past. Perhaps in the end he and his great contemporary rival Linnaeus had not seen too differently—though Buffon had seen farther and been bolder about what he saw.

The count died in 1788, ten years after Linnaeus, although both had been born in the same year. It was a good time to go. The next year his son was to perish with other aristocrats in the fury of the Terror, proudly and reproachfully saying as he waited on the scaffold, 'Citizens, my name is Buffon.' It was the end of an age. Buffon like Linnaeus had had a world reputation, had had his specimens passed graciously through warring fleets, had corresponded with Franklin, had been one of the leading figures in a great country at the height of its intellectual powers.

It is a great pity that his ideas were scattered and diffused throughout the vast body of his *Natural History* with its accounts of individual animals. Not only did this concealment make his interpretation difficult, but it lessened the impact of his evolutionary ideas. If he had been able to present his thesis in a single organized volume, it is possible that he himself might have argued his points more cogently and perhaps seen more fully the directions of his thought. For one important idea was still lacking: Buffon had never wrestled satisfactorily with the mechanism of change. He seemed, at times, quite conscious of the value of selection in breeding experiments, and this, with his full recognition of individual variation, implies something very close to Darwin's later theories. Actually, however, Buffon never seems to have been able to get from artificial to natural selection. Instead he suggests 'climate' as a leading factor in 'degeneration'. However, almost everything necessary to originate a theory of natural selection existed in Buffon. It needed only to be brought together and removed from the protective ecclesiastical coloration which the exigencies of his time demanded.

Write a succinct criticism of Eiseley's approach, making as many different points as you can.

I would want to make the following points:

Eiseley's account has some good features, including his useful identification of ideas which were to feature prominently in Darwin's synthesis. At the same time he falls into almost every trap outlined above. His approach smacks of the whig historian who is preoccupied with the search for precursors: Buffon managed to mention *every significant ingredient*; he did not *quite* manage to put them together; he *foreshadowed* this, he *anticipated* that, he had a *premonition* of something else. The very fact that what Eiseley considers to be seminal ideas were so scattered and diffused ought to have suggested to him, not that the fact should be lamented ('it is a great pity . . .'), but that the ideas were not as significant for Buffon as they are for Eiseley himself: to suggest that if only Buffon had rearranged them he would have seen 'more fully the direction of his thought' is a prime example of writing history backwards. To attach so much importance to a belief in degeneration is to forget that change in species was, for Buffon, the exception and not the rule. Eiseley also overlooks the general point that some of Buffon's *other* ideas may have militated against a systematic theory of transformation. As Wilkie emphasizes, in Buffon's outlook a theory of spontaneous generation could have been adequate as an explanation for the origin of species without having to postulate a theory of transformation at all. Nor would Buffon's criterion for a common ancestry admit progressive evolution: mules were—or so he thought for some time—sterile as well as stupid. The real weakness of Eiseley's reconstruction is associated with his failure to ask what questions were uppermost in Buffon's mind, what were *his* problems, what were *his* intentions? As Wilkie explains, his vision in compiling the massive *Histoire Naturelle* embraced everything from plants and animals to mineralogy and anthropology, with the intention of exposing the system builders. This is particularly true of the controversial article on the ass, where Buffon's intentions were directed against systematic classification and not towards 'evolution'. To Maupertuis' observations on the possibility of evolution by chance variation, Buffon replied with scepticism: without evidence from the hybridization of widely dissimilar forms, there was insufficient empirical support. Finally, despite his caring nothing for revelation, and despite his having been irritated by the theologians of the Sorbonne, it does not follow that his rejection of evolution was ironic or determined by religious diplomacy. It is therefore misleading of Eiseley to suggest that 'almost everything necessary to originate a theory of natural selection existed in Buffon. It needed only to be brought together and removed from the protective eccelesiastical coloration. . . .'

Buffon's concept of species has been and will continue to be a source of controversy among historians, but you should now be in a position to see how simplistic it would be to portray him as a precursor of Darwin who was stopped in his tracks by the batteries of the Sorbonne.

If you now read pp. 259–61 of Wilkie's essay, you will discover a further reason why this picture could be simplistic. What is the reason?

From his knowledge of the distribution of animals, and particularly the existence of species peculiar to the New World, Buffon sometimes found it hard to believe that all those peculiar species had ancestors in common with species in the Old World. Even when more favourably disposed to the idea, his hesitation in accepting a whole-hearted theory of evolution sprang from considerations that were not theologically conditioned.

To conclude that Buffon was *not* a precursor of Darwin in any meaningful sense, is not of course to condemn him. Rather the opposite, if anything, as Wilkie himself insists (see p. 262). In fact you should be in a better position to understand the real originality of his thought. Within a natural 'family' a species type or *première souche* could undergo gradual change, but never so drastic a change as would prevent successful mating with the original members of the family. Each family was distinct, so there was no notion of common descent transcending family barriers. But Buffon was able to effect a conspicuous reduction in the number of species. All domestic cattle could be reduced to two species, the buffalo and the cow: all known sheep could be conceived as degenerations of a primitive wild type. In this respect Buffon was breaking new ground. As Paul Farber has recently declared, his 'popularity in his own day suggests that his contemporaries appreciated his role better than posterity'—appreciated him for having developed a model of the history of life which, though not evolutionary in the modern sense, could still be construed as an alternative to the Genesis account.[1]

3.3 Cuvier and transformation

If our understanding of Buffon has been bedevilled by crude historiography, the same may be said of the foremost naturalist of the next generation, Georges Cuvier, who dominated the world of comparative anatomy and The Paris Academy for many years. As an influential critic of evolutionary theories, Cuvier presents a problem to those historians who habitually presuppose warfare between enlightened evolutionary scientists and the religiously inclined. The warfare thesis (see Unit 1) when applied to Buffon generated an image of a convinced evolutionist stalled by religious censure; when applied to Cuvier it generated an image of a great man blind to his *own* religious preconceptions which *must* have twisted his reading of nature. In the extraordinary words of A. D. White:

> For the man who now took up the warfare, avowedly for science but unconsciously for theology, was the foremost naturalist then living—Cuvier.

Your task in this section is simply to write a succinct critique of this remark. Read first of all the extract from William Coleman's biography of Cuvier in *Anthology II*, pp. 224-37. With this as a basis can you indicate the several respects in which White's judgement is defective?

My critique of White's remark would look something like this:

To speak of 'warfare' at all is anachronistic since one of Cuvier's motives was to demonstrate the concordance between his science and his religion. White's categories are crude in the extreme because they exclude the possibility of scientific arguments against 'evolution', and of theological arguments for it. Moreover, Cuvier's own arguments against the transformation of species were primarily scientific and philosophical; his religious beliefs played but a secondary role. It was his interpretation of the *principle of correlation* that most militated against his acceptance of evolution. Each species was the result of a harmonious combination of structures—the function of each part in perfect correlation with the function of others. In accordance with this teleological principle, any isolated variation in the construction of a part would lead to extinction, rather than improvement: webbed feet would not befit an eagle! In order to survive, an organism would have to change en bloc, and to Cuvier, as to most of his successors, such an enormous mutation was inconceivable. In explaining Cuvier's resistance to transformation one would also have to discuss what may have been a psychological aversion to systems which postulated change; and, as a taxonomist, his criteria for the classification of species required a stability which schemes of transformation could not provide. Though his religious commitment may have endorsed the correlation principle, he was neither a 'doctrinaire Biblical zoologist' nor a typical proponent of the argument from design. His religious convictions required no scientific endorsement, and, whatever his political motives, he could evaluate the transformation hypothesis in terms which, to him, were entirely scientific.

Figure 4 Cuvier, lithograph by Delpech from a sketch by Maurin (Mansell Collection).

[1] P. L. Farber, 'Buffon and the Concept of Species', *op. cit.*, p. 283.

As Coleman stresses, it was not until Darwin that the correlation of parts, and the adaptation of each species to its environment, could be reduced to the agency of 'natural' forces. It should not be surprising then that many contemporaries of Cuvier should feel that his rejection of species transformation was both incisive and definitive. Henceforward those who in France wished to side with Lamarck had the reputation of Cuvier to contend with. In England, as you will see in the next section, they were soon to have that of Lyell. Before you proceed, however, you ought to ask yourself what you have learned from this section. Could you restate the traps involved in the common quest for Darwin's precursors? Could you use the examples of Buffon and Cuvier to expose the deficiencies of the warfare thesis?

4 Lyell and Lamarck

4.1 The development hypothesis of Lamarck

So far in this unit we have focused much of our attention on problems which constituted stumbling blocks to the acceptance of evolutionary theories. This is not from a biased reading of the pre-Darwinian period, but is intended as a reflection of the many scientific difficulties that had to be overcome before critical minds could accept a thesis of common descent. But it is obviously important that you study at least one variation on the evolutionary theme, if only to appreciate what it was that Cuvier had rejected. During the fifty years from 1770 to 1820 Jean Baptiste Lamarck devoted himself to the study of natural history, and in doing so proposed a mechanism for evolutionary change, which, though often compared unfavourably with that of Darwin, was sophisticated enough to torment conservative minds with the possibility of a long and low genealogy, and sophisticated enough for Darwin himself to use parts of it in later editions of the *Origin* when the sufficiency of 'natural selection' was called into doubt.

Lamarck has been described as the foremost taxonomist of his day and the principal founder of biology as a unique discipline with methodological problems of its own. By 1815 he reached the conclusion that the most distinctive feature of living things was their power to respond to internal and external stimuli. Nor did Lamarck hesitate to apply this insight to man himself, who, as part of nature, had great potential for change. Development towards greater complexity was nature's norm.

You will find an exposition of Lamarck's theory in the essay by Professor Wilkie to which you have already been referred. Read pp. 238–45 and p. 263 to the end. You should then be able to answer the following questions.

1 Wilkie distinguishes between what he calls the 'general theory of evolution' and the Lamarckian theory. What is the significance of this distinction from the historian's point of view? Are there dangers in talking about a 'general theory' at all?
2 How would you illustrate the proposition that Lamarck was a deist?
3 What do you understand by the doctrine of 'the inheritance of acquired characteristics?'
4 Which of the following propositions appear to be true and which false?

- **[a] Lamarck's theory of evolution remained essentially the same from the first version in 1801 to the final version in 1815.**
- **[b] Lamarck could never renounce the idea of a single chain of being.**
- **[c] The principle of the inheritance of acquired characteristics is used by Lamarck merely to explain observed deviations from a regular progression of living forms.**
- **[d] By 'nature's plan of campaign in the production of animals' Lamarck simply meant that it was possible for living organisms to become progressively more complicated.**
- **[e] Lamarck sincerely believed that in directing the process of evolution God was aiming directly at the production of man.**
- **[f] Lamarck felt constrained to couch his theory in religious terms.**
- **[g] There is an inconsistency in Lamarck's theory since, if every creature has had an innate tendency to become more complex, there should no longer be living representatives of the inferior links in the chain.**

1 It was in principle possible to believe in the general 'fact' of evolution—in the progressive transformation of species—without subscribing to any one particular mechanism. However, the fact that Comte rejected transformation could be used as evidence to suggest that neither Lamarck's mechanism nor that of anybody else had proved convincing enough to sustain any general theory. Furthermore, Darwin's subsequent stress on the mechanism of natural selection suggests that in order to stand up to scrutiny any *general* theory had to be more than a plausible idea. To introduce the notion of a general theory at all is potentially dangerous, since it could straitjacket discussion with its implication that Darwin's predecessors were perhaps all seeking a mechanism for *one and the same* theory. Wilkie himself may have underestimated this danger!

Figure 5 Lamarck, engraving by Ambroise Tardieu (Mansell Collection).

2 Lamarck's deism could be illustrated with reference to:

(a) his commitment to the chains of being which conditioned his adoption of a linear tendency in evolution long after he had realized that a linear scheme of classification was untenable;

(b) his conviction that God's creation was rationally ordered and intelligible—a conviction which made it difficult for him to accept that species had become extinct.

(c) his optimism which precluded any constructive appreciation of the struggle for existence.

3 The doctrine of the inheritance of acquired characteristics could be said to embrace the last three of the four laws cited on p. 268; i.e. the production of a new organ results from a new need; the development reached by organs is directly proportional to the extent to which they are used; everything acquired by the individual is transmitted to its offspring. A strict answer to the question would mention the last proposition alone.

4 (a) False—since the inheritance of acquired characteristics was the primary mechanism in 1801, but subordinate in 1809.

(b) False—since Lamarck adopted two independent lines, one for plants, one for animals; and in 1815 entertained the possibility of two animal series.

(c) True—in the mature statements of his theory.

(d) False—the tendency towards greater complexity was not merely possible; it was constantly at work, effected by a primal cause.

(e) True—in so far as it is ever possible to judge with certainty.

(f) False—since Lamarck acknowledged that in his own country the prevalent spirit of the times was naturalistic and pantheistic if not atheistic.

(g) Possibly true, according to Wilkie, though a belief in a continuous production line of the humblest creatures—by some process of spontaneous generation—would remove the alleged inconsistency.

From this exercise you should have a fair understanding of the essentials of Lamarck's theory. There are of course problems of interpretation, and his contentions have often been misunderstood. One problem that comes to mind immediately concerns the allegation of inconsistency made by Wilkie who claims that if Lamarck were obliged to admit spontaneous generation his theory would be robbed of simplicity and elegance. Is this really the case? Or does Lamarck's admission of spontaneous generation help us to see more clearly the sort of evolutionary theory he was really propounding? Study the extract below and then answer the question which follows:

> I do not doubt that . . . water is the true cradle of the entire animal kingdom. We still see, in fact, that the least perfect animals, and they are the most numerous, live only in water . . . that it is exclusively in water or very moist places that nature achieved and still achieves in favourable conditions those direct or spontaneous generations which bring into existence the most simple organized animalcules, whence all other animals have sprung in turn. . . .[1]

[1] Cited by E. Mayr, 'Lamarck Revisited', *Journal of the History of Biology*, 1972, **5**, 1, 60.

Does this passage suggest that progressive transformation has been a once and once only process, or does it suggest that the process of development may be repeating itself continuously as each batch of the lowest organisms appears at the foot of the escalator?

This is a difficult question, but it is clear from the passage that Lamarck does believe in repeated spontaneous generation—generation that is still taking place today. Since all organisms have an innate tendency to become more complex, this must surely mean that the whole evolutionary process could be continually repeating itself? The humblest organism, spontaneously generated today, could, in principle, run the same course as those which emerged thousands of years ago—providing environmental conditions induced the same responses.

If this is the correct interpretation of Lamarck, it only confirms once again how dangerous it is to speak of him as an evolutionist in anything like the sense in which that word may be applied to Darwin. For, in Darwin's theory of common descent, the evolutionary process was unique, irreversible, non-linear, and utterly unrepeatable: species once extinct were extinct for ever.

As Ernst Mayr has recently reminded us, two more problems at least have recurred in the literature: did Lamarck really believe that changes in the environment *directly* induced change in a species; and secondly, did Lamarck actually commit himself to the assertion that an organism could change and adapt itself by its own volition? Could a giraffe, by the very act of *desiring* a longer neck, in the course of time acquire one? A passage which may throw some light on these further problems of interpretation is reproduced below. Study it carefully and then answer the question which follows:

> I must explain at this point what I mean by the statement: *Circumstances* (*circonstances*) *influence the form and organization of animals*—that is to say, when these circumstances become quite different they alter both the form and organization through proportionate modifications during the course of time.
>
> Certainly, if these statements were taken literally I could be accused of error; for, whatever environmental circumstances may do, they do not directly produce any change whatsoever in the shape and organization of animals.
>
> However, major changes in the environmental circumstances of animals bring about changes in their needs, and such changes in turn necessarily bring about changes in their actions. Now, if their new needs become constant or especially long-lived, the animals then assume new *habits*, which are as long-lasting as the needs which produced them. This is easy to demonstrate, and indeed no explanation is required for it to be perceived.
>
> It is, then, evident that a great change in the circumstances having become constant for a race of animals, leads these animals to new habits.
>
> Now, if new circumstances, having become permanent for a race of animals, have produced new *habits* in these animals—that is, have led them to new actions which have become habitual—the result will be the use of such parts that are preferred to others, and in certain cases, the total neglect (lack of use) of a part which has become useless.
>
> By citing well-know substantiating facts, we shall see presently that new needs—having made a part necessary—really have given birth to that part by virtue of repeated efforts; and that subsequently its constant use gradually has strengthened, developed, and finally, enlarged it considerably.[1]

What light does this extract throw on the two problems of interpretation mentioned above?

In the first place Lamarck admits that if he were taken literally he could be accused of implying the *direct* influence of the environment on the form and organization of

[1] McKinney, *op. cit.*, pp. 9–10.

animals. But he immediately adds that this is not what he really believes. In the second place it is not entirely clear whether Lamarck is presupposing the causal agency of volition or not. His emphasis falls on new *needs* and new *habits*, rather than new *desires*. At the same time he does say quite explicitly that a new part or organ can be generated 'by virtue of repeated efforts'. But the problem of interpretation still remains, since it is not clear whether these repeated efforts are consciously or unconsciously made.

It is worth noting that for those who read Lamarck in the original French this same problem was exacerbated by a difficulty in the translation of the world *besoin*. In the translation you have just read, *besoin* is correctly rendered as 'need', but if it was translated as 'desire' or 'want' — as it was by Cuvier and sometimes by English interpreters—the case for the 'volitional' interpretation looked quite strong. Lamarck did in fact clarify his position when he argued that, as a consequence of environmental change, animals *felt* certain needs, and it was each 'felt need' which, by stirring their inner consciousness (*sentiment intérieur*), caused various fluids and forces to be directed towards the appropriate part of the body.[1] In other words, felt needs rather than direct exertion of the will were the propellants for organic modification. Nevertheless, a mere acquaintance with Lamarck's writings, which is all that most Englishmen could aspire to before 1830, would leave his work wide open to misunderstanding.

This is an important point because the very fact that Lamarck was French, the very fact that as an evolutionist he was bound to be associated with the 'atheism' of Laplace whose nebular hypothesis had promoted a scheme of cosmic evolution, meant that in England, scholars who were still sensitive to the horrors of revolution were almost bound to harbour suspicions. If Lamarck could be caricatured so much the better. It does not matter whether Paley had Lamarck in mind when he satirized a system 'which has lately been brought forward, and with much ingenuity.' The fact is that Lamarck would have been dealt the same treatment:

> A piece of animated matter . . . that was endued with a propensity to *fly*, though ever so shapeless, though no other we will suppose than a round ball, to begin with, would, in a course of ages, if not in a million of years, perhaps in a hundred millions of years (for our theorists, having eternity to dispose of, are never sparing in time), acquire *wings*. The same tendency to locomotion in an aquatic animal, or rather in an animated lump which might happen to be surrounded by water, would end in the production of fins: in a living substance, confined to the solid earth, would put out *legs* and *feet*; or, if it took a different turn, would break the body into ringlets, and conclude by *crawling* upon the ground.[2]

In England caricature was possibly not displaced by critical analysis until Lyell examined Lamarck's theory in the *Principles of Geology*, and it was from Lyell's paraphrase of Lamarck that his contemporaries could henceforward draw their own conclusions. Lyell was anxious that Lamarck should not be misunderstood—indeed he later reproached Darwin for attributing to him the agency of volition—but he was equally anxious that Lamarck should not be believed.

4.2 Lyell versus Lamarck

The historian's task in discussing the pre-Darwinian period would have been made much easier had Lyell defended Lamarck. One might then have no inhibitions about portraying Lyell as a typical precursor of Darwin, who, having eliminated divine intervention from the history of inorganic phenomena, was prepared to do the same for the history of life. The trouble is that he was not. Lamarck's theory of continuous, progressive transformation was reviewed but rejected. The authority of Lyell was to be squarely behind the more conservative belief that pairs of each species had been

[1] J. C. Greene, *The Death of Adam*, Mentor, New York, 1961, p. 163.
[2] W. Paley, *Natural Theology*, London, 1818, p. 356.

created at different times and different places, their offspring dispersing into localized areas as a consequence of the gradual erection and removal of natural barriers. We shall, accordingly, have to ask why this was. On what grounds did Lyell reject Lamarck?

In general terms Lyell recoiled because he still believed that species were real and distinct, and because he felt that Lamarck's theory contravened what he understood by the uniformity of nature. Of the constancy of nature he could write: 'if there once appears ground of reasonable doubt, in regard to the constancy of species, the amount of transformation which they are capable of undergoing may seem to resolve into a mere question of the quantity of time assigned to the past duration of animate existence.'[1] And of the notion of progressive change he could say quite specifically that it constituted a threat to 'the assumption of uniformity in the order of nature.'[2]

Why it should do so may not be immediately obvious, but from your knowledge of Lyell can you say why he should argue the point?

Lyell believed that the fossil record was too imperfect to justify belief in any *progression* during the history of life. There was, he claimed, a strict uniformity about the pattern of life which precluded progressive transformation oι the kind postulated by Lamarck.

This takes us some way towards understanding Lyell's position, but we need to look more closely at the details of his arguments in order to gain any real insight into his strategy and motivation.

The fact is that Lamarck had in a very real sense fulfilled the uniformity of nature and not transgressed it. There was no reason in principle why there should not be uniform development, but for Lyell there was a good reason in fact. Against Lamarck he protested that although habitual exercise might conceivably modify organs and structures that were already in existence, the French zoologist had not provided a convincing explanation for their *origin*. In distinguishing between modification and origination, Lyell had put his finger on one of the most serious stumbling blocks to the acceptance of Lamarck's hypothesis. One might point to the immense variety of dogs produced by domestic breeding, but this could not provide 'even the semblence of a foundation for the theory of Lamarck, respecting the growth of new organs and the gradual obliteration of others.' Why not? Invoking the authority of Cuvier, Lyell insisted that 'in all these varieties of the dog . . . the relation of the bones with each other remain essentially the same.'[3] In the case of oxen, cats and crocodiles it was not only the relation of the bones that remained the same. When mummified specimens dating from c. 3000–2000 B.C. were retrieved on Napoleon's expedition to Egypt (1798–99), it was discovered that in five thousand years cats had not ceased to be cats. So here was an easy debating point. This creature 'has been carried over the whole earth, and, within the last three centuries, has been naturalized in every part of the new world . . . yet it has scarcely undergone any perceptible mutation and is still the same animal which was held sacred by the Egyptians.'[4] Five thousand years had disclosed no direct evidence for transformation.

Defending the stability of species, Lyell had to do justice to reports of successful experiments on the hybridization of plants—notably those of Kolreuter who, by the successive fertilization of hybrid generations, had changed one *Nicotania* species into another. Was this not direct evidence for the possibility of transformation? Not at all, Lyell replied. Hybridization might redistribute or exclude the parental characteristics, but it could never generate *new* structures. Hybrids might not all be sterile but there were few which were sufficiently prolific ever to survive in the wild state. Turning his attention to the distribution of species, Lyell was struck, as Buffon before him, by the restriction of groups of distinct species to regions isolated from the rest of the earth

[1] W. Coleman, 'Lyell and the "Reality" of Species: 1830–33', *Isis*, 1962, **53**, 326.
[2] C. Lyell, *Principles of Geology*, 1830–33, vol. I, chapter 9, p. 144.
[3] Lyell, *op. cit.*, vol. II, chapter 2, p. 27.
[4] Lyell, *op. cit.*, vol. II, chapter 2, p. 30.

by natural barriers. What should one infer from such a pattern of localized distribution?

> This congregating, in a small space, of many peculiar species, would give an appearance of *centres* or *foci* of creation, as they have been termed, as if there were favourite points where the creative energy has been in greater action than in others.

This was the appearance. But the reality, according to Lyell, was otherwise. Species had been created at points distributed evenly, 'uniformly', through space and time.[1]

Lyell preferred to speak of 'creative energy' rather than tolerate any process of progressive modification. Why did he marshal his arguments in this way? What presuppositions were directing his thought? What was his motivation? These are difficult questions but the three extracts reproduced below may help you to gain further insight into Lyell's position. Study each one in turn and then answer the question which follows.

> . . . even in the oldest rocks which contain organic remains, some genera of marine animals are recognized, of which species still exist in our seas, and these are repeated at different intervals in all the intermediate groups of strata, attesting that, amidst the great variety of revolutions of which the earth's surface has been the theatre, there has never been a departure from the conditions necessary for the existence of certain unaltered types of organization.

> . . . [with the selective breeding of plants] we soon reach certain limits, beyond which we are unable to cause the individuals, descending from the same stock to vary: while, on the other hand, it is easy to show that these extraordinary varieties could seldom arise, and could never be perpetuated in a wild state for many generations, under any imaginable combination of accidents. They may be regarded as extreme cases brought about by human interference. . . . The propagation of a plant by buds or grafts, and by cuttings, is obviously a mode which nature does not employ.

> If the organization of man were such as would confer a decided pre-eminence upon him, even if he were deprived of his reasoning powers, and provided only with such instincts as are possessed by the lower animals, he might then be supposed to be a link in a progressive chain . . . [but he who believes in the transmutation of species] renounces his belief in the high genealogy of his species, and looks forward, as if in compensation, to the future perfectibility of man in his physical, intellectual, and moral attributes.

What light does each passage throw on the reasons for Lyell's opposition to progressive transmutation?

1
Lyell is clearly impressed by the fact that certain marine genera had not altered despite the ravages of time. By analogy one could proceed to argue that no unilinear development of organic form had necessarily occurred.

2
Lyell's argument is directed against a supposed analogy between human selection and the hand of nature. The contrast between 'human interference' and the 'wild state' is so great that analogies drawn from domestic breeding cannot sustain any theory of transformation.

3
The pre-eminence of man, since it rests not on his physical organization but on his powers of reason, makes it impossible to believe that he could be the product of a progressive chain.

Figure 6 Charles Lyell, sketch by Maguire (Mansell Collection).

---

[1] M. J. S. Rudwick, 'The Strategy of Lyell's *Principles of Geology*', *Isis*, 1970, **61**, 20.

Each of these arguments raises a host of questions, but in the last analysis they were probably the least superficial from Lyell's point of view. His stress on the 'conditions necessary' for unaltered life suggests perhaps that, providing appropriate conditions were re-established, creatures similar to those once extinct might reappear? His attack on the analogy between human selection and 'natural selection' is particularly interesting since this was to be a main plank in Darwin's argument. But it is Lyell's respect for the high genealogy of man which could provide a *motive* for his attack on the thesis of progressive development. This last interpretation of Lyell has recently been advanced and refined by Michael Bartholomew, who has suggested that it was Lyell's reading of Lamarck in 1827 which determined him to resist all notion of progress in the fossil record precisely because the uniqueness of man would otherwise be put in jeopardy.[1] Any progressive trend in the history of life could always be extrapolated to include man himself, and for Lyell that was a distasteful prospect. Whether his belief in the dignity of man was primarily a religious belief it is difficult to judge. There is evidence that Lyell was unitarian by persuasion; living in London he would quite regularly attend the chapel in Little Portland Street. And if his unitarianism was a derivative of deism rather than a vestigial Christianity, there might be a significant correlation between the a-historical basis of his religion and the a-historical motifs of his geology (compare Dr Richardson's thesis in Unit 14).

That Lyell so resolutely turned his face against the hypothesis of progressive development should not blind us to the fact that his own sustained account of the effect of geological forces on the distribution, dispersal and extinction of species prepared the ground more thoroughly than any predecessor for the kind of questions that Darwin and Wallace were to ask. Indeed it was possible, as Wallace later did, to read the *Principles of Geology* and to come to precisely the opposite conclusion concerning the constancy of species. But that is to anticipate. Between Lyell and the published theses of Darwin and Wallace, another scheme of progressive development made its appearance. The *Vestiges* of Robert Chambers, published anonymously in 1844, boldly argued for a law of development that would explain the emergence of man from beast. Frequently criticized for its amateurishness it did as much to discredit as establish the possibility of transformation, but in the outcry that followed the sting was withdrawn from the opposition that was later to be aimed at Darwin.

Before you proceed, however, you might like to test your comprehension of Lyell and Lamarck by specifying as many points of contrast between them as you can. Check your answers by quickly re-reading this section.

5 Robert Chambers and the Natural History of Creation

5.1 Chambers and Sedgwick

When *Vestiges* appeared there was well nigh a consensus of opinion among professional geologists and naturalists: a fool had stepped in where angels had feared to tread. The pride of the professional throbs through the reviews, and in the case of Adam Sedgwick, Professor of Geology at Cambridge, hysteria lapsed into quotations from Lear: 'Give me an ounce of civet, good apothecary, to sweeten my imagination.' You will find an introduction to Chambers and to Sedgwick's response in *Gillispie* (set book), chapter 6. Read the first three sections (pp. 149–70), and then answer the questions below.

[1] M. Bartholomew, 'Lyell and Evolution', *British Journal for the History of Science*, 1973, 6, 261.

1 What reason did Chambers give for writing the *Vestiges*?
2 To which basic analogy did Chambers appeal to justify a law of progressive development?
3 What pronouncements about *man* did Chambers make that were liable to be offensive?
4 State whether the following propositions are true or false:
[a] The author of *Vestiges* had a profound contempt for natural theology.
[b] The progressive development of organic life was the central thesis of Chambers' book.
[c] Chambers' theology of nature was deistic.
[d] There was no contradiction between his theory and the Genesis account of creation.
5 Can you list the accusations levelled against the book?
6 On page 167, Gillispie quotes a long passage from Sedgwick and suggests that it was not far removed from the alleged 'materialism' of Chambers. Compare that passage with those cited on page 153. Can you detect a difference in emphasis in Sedgwick's discussion?

Figure 7 Robert Chambers, pencil sketch (Mansell Collection).

1 Chambers claimed that scientists had become too specialized: they could not see the wood for the trees (pp. 161–2).

2 You could answer this question in at least two ways. On the one hand, Chambers appealed to an analogy with the *law* of gravitation to justify the proposition that in the organic sphere one *law* of development obtained (p. 153). On the other hand, he appealed to an analogy with a developing embryo to justify the proposition that his law was one of *development* (p. 156).

3 Chambers' remark that 'man, considered zoologically, and without regard to the distinct character assigned to him by theology, simply takes his place as the type of all types of the animal kingdom'—this remark could be construed as an invitation to disregard the claims of theology. He implies an extrapolation of organic development which might lead to species superior to man. He asserts that there is nothing unique about the *mind* of man (p. 157); and announces to the world that all distinction between physical and moral is annulled. With a large enough sample man was merely a mathematical problem: he could be reduced to a moral statistic.

4 (a) False—Far from finding natural theology contemptible, Chambers was motivated by the desire to establish God's beneficent rule, and he could say that the only value of his book lay in its religious import (p. 159).
(b) False—The central thesis was the dominion of natural law, which by its very unity testified to its divine source (p. 153).
(c) This is difficult to answer. It would (and did) appear so. At the same time he did see the 'Eternal One . . . present in all things' (p. 155).
(d) True for Chambers, but false for Sedgwick!

5 It was said to contradict the Biblical affirmation that man was made in the image of God; to destroy all distinction between physical and moral; to promote a degrading materialism; to offend all decent sense of propriety; to undermine the moral foundations of society (and consequently to be dangerous for the working man); and to be the work of an ignorant and unscientific mind which could accept absurdities like spontaneous generation.

6 Despite the superficial, though striking, resemblance, Sedgwick refers to the adaptation of means to ends—those immediate designs in nature from which the *goodness* of God may be inferred. Similarly he does not believe that material law could be an adequate instrument of God's Will. According to Sedgwick God is not merely present in all things but present in such a way that He can be recognized as a *personal* God whose will may affect every atom.

Figure 8 Adam Sedgwick (Radio Times Hulton Picture Library).

Of Chambers' theory Gillispie writes that 'no scientist could see any meaning in the only mechanism suggested,' and he proceeds to quote the difficult passage about the discharge of globules (p. 156). However, to dismiss Chambers in this way is an easy option for the historian. If one makes the mistake of supposing that Chambers was offering a mechanism for an essentially *Darwinian* theory, then obviously he suffers by comparison with the professional naturalist. But, as Hodge has insisted, the mistake is a serious one.

5.2 Chambers and Darwin

To portray Chambers as a 'precursor' of Darwin is particularly misleading when one remembers that by 1844, when *Vestiges* appeared, Darwin had already prepared the first sketch of his theory. It is even more so when one realizes that Chambers' whole exercise was different in its conception and in its execution. When we understand Chambers' problems and intentions we shall be in a better position to understand his mechanism. The object and the scope of *Vestiges* were altogether different from the task which Darwin had set himself. Whereas Darwin was restricting himself to the biological problem of speciation, Chambers was searching for a grand cosmic scheme which would convince every man that events in the natural world were the consequence of invariable laws—laws which revealed that God's intent was universal happiness. Designed by an amateur for popular consumption, his book presented so sweeping a canvas that vistas opened in all directions, even as far as the planets which were populated by the same set of laws.

Given Chambers' intention of reducing the universe to a few well chosen laws, his problem was that of imposing law on the history of life. This was no mean undertaking, even for a Scotsman, and he confessed himself that he had been puzzled by the origin of the more complex living things. Spontaneous generation might work for the most primitive organisms but what of mammals which required internal gestation? There could be but one answer: 'gradual evolution of higher from lower, of complicated from simple, of special from general, all in unvarying order, and therefore all natural, although all of divine ordination.'[1] The law had to be a law of development, and the only plausible analogy was that of the developing foetus. As Hodge implies, this is the key to understanding Chambers since the analogy with the embryo was not merely illustrative but constitutive. The process of reproduction could provide the mechanism for the progressive development of life. For us the connection may be far from clear but for Chambers himself the crucial consideration was the sequence of stages which a mammalian foetus appeared to pass through prior to birth. In its early stages it resembled the early foetus of a fish, at a later stage the early foetus of a reptile, then bird, then mammal. According to Chambers the foetus of fish, reptile, bird and mammal all pass through the same initial stages before each diverges along its own line of development. With this assumption a flash of imagination gave Chambers his answer. In Hodge's paraphrase: to advance from one class to another, an embryo, instead of diverging from the straight line at the usual point, merely continues to the next so that the progeny will not be a fish say, but a reptile.[2] It was only necessary for Chambers to suggest conditions under which abnormal gestation might occur in order to complete his mechanism. Any factors which might prolong the gestation period, for example, would lead to an advance of type. 'Whenever oats sown at the usual time are kept cropped down during the summer and autumn and allowed to remain over the winter a thin crop of rye is the harvest at the close of the ensuing summer.'

Pause for a moment to compare this testimony of Chambers with Gillispie's account of it on page 156. What criticism of Gillispie would you make?

Apart from the cynicism of Gillispie's account, which is not without justification, he does not expound the real significance of the alleged observation, which is that oats, under *protracted gestation*, could turn into rye. Chambers' mechanism would not have appeared quite as meaningless as Gillispie suggests.

That it was not meaningless does not of course mean that it was correct! When Chambers tried to correlate each stage of foetal advance with the ascent of rocks through time his analogies must have seemed contrived and fragile (see Table 2). But there is

[1] J. Hodge, 'The Universal Gestation of Nature', *ibid.*, 138.
[2] Hodge, *ibid.*, 143.

no doubt that they enabled him to explain what he took to be the four crucial facts: a unity of organization among diverse types; the resemblance between present and fossil species; the existence of rudimentary organs, and the historical dimension of life as inferred from the study of palaeontology. What he did not do was to explain these 'facts' in a manner remotely reminiscent of Darwin. Though Wallace was prepared to say that *Vestiges* had exerted a major influence on his thinking, both he and Darwin very properly insisted that Chambers had not begun to touch the question of how new, distinct, species could come into existence. To see Chambers as a 'precursor' also obscures the fact that whereas Darwin was to explain the origin of species with reference to branching lines of development from a common source, Chambers' analogies presupposed linear development from independent sources. The contrast between them has been very nicely brought out by their totally different reactions to data which Darwin had collected on the *Beagle* voyage. The absence of mammals on the Galapagos was interpreted by Chambers as a vindication of his theory: the requisite time had not yet elapsed for life to reach the mammal stage. But if there had been mammals *without* reptiles that would have constituted 'one decided fact against the development theory.'[1] The interpretations of Darwin and Wallace were to be altogether different. Whereas Chambers required a young Galapagos to explain unfinished development, Wallace demanded a Galapagos sufficiently ancient to ensure long isolation from the mainland—isolation which would help to explain the differences between mainland and island species. Since Darwin believed that all

TABLE 2 *The vegetable and animal kingdom**

ASCENDING SERIES OF ROCKS	FOETAL HUMAN BRAIN RESEMBLES, IN
1 Gneiss and Mica Slate system	
2 Clay Slate and Grawacke system	
3 Silurian system	1st month, that of an avertebrated animal;
4 Old Red Sandstone	
5 Carboniferous formation	2nd month, that of a fish;
6 New Red Sandstone	3rd month, that of a turtle;
	4th month, that of a bird;
7 Oolite	
8 Cretaceous formation	
9 Lower Eocene	5th month, that of a rodent; 6th month, that of a ruminant;
10 Miocene	7th month, that of a digitigrade animal;
11 Pliocene	8th month, that of the quadrumana;
12 Superficial deposits	9th month, attains full human character.

* From Robert Chambers, *Vestiges of the Natural History of Creation*, Leicester University Press Reprint, 227.

[1] Hodge, *ibid.*, 148.

mammals had descended from a common ancestor, their absence on the Galapagos was simply due to the absence of successful migration.

Darwin probably learned little from *Vestiges*—except perhaps to avoid words like 'higher' and 'lower' when talking about successive species. But from the reviews, and particularly that of Sedgwick, he gleaned the full measure of the opposition which had still to be overcome if any theory of transmutation was to survive. It has in fact been argued that the reason why Darwin read the sixth edition (1847) so carefully was to see how successfully Chambers had dealt with his critics.[1] Among them, Sedgwick may have been the most vituperative, but Hugh Miller perhaps more rewarding from the historian's point of view. A fellow Scot, sharing with Chambers both a passion for geology and popular education, he quietly but effectively dismissed *Vestiges* as an 'ingenious, well-written book, unfortunate in its facts, and not always very happy in its reasonings.'

5.3 Chambers and Miller

You will find a sympathetic treatment of Hugh Miller in *Gillispie* (set book), pp. 170–83. Study Gillispie's discussion and then answer the questions below.

Figure 9 Hugh Miller, engraved frontispiece of Hugh Miller, My Schools and Schoolmasters, *1872.*

1 Why are the works of Miller worth the historian's attention?
2 How could Miller regard the popularization of science as a religious duty?
3 Does it make sense to describe Miller as a 'progressive creationist'?
4 What was his most deeply felt objection to the development hypothesis?
5 'The development hypothesis could be reconciled with the Christian faith simply by admitting that God had inserted an immortal soul at an apposite stage in the process.' Why did Miller reject this proposition?
6 Did Miller believe that it was impossible for science to clash with the Bible?
7 What was Baden Powell's objection to Miller's writings on the relationship between science and religion?

---

1 Basically because they sold so many copies, even when compared with Lyell and Chambers: and because they give us insight into the reactions of an evangelical Christian who was also a respected scientist.

2 Miller undoubtedly believed that the sciences revealed the wonder of God's creation; that they were of great educational value to those who, like himself, were self-taught; that geology was particularly well suited to ennoble the mind; and that the layman must be informed in order that he should not be deceived by the corrupting influences of pseudo-science.

3 If by 'progressive creationist' we mean someone who believes that God's creative power has been more greatly exercised at the beginning of each new epoch, then Miller would seem to fit the description. He clearly believed in successive 'dynasties' each more advanced than the previous and eventually culminating in a terminal dynasty when the 'progress Godwards' would reach its destiny. Miller did, however, differ from some progressive creationists when he insisted that within the great geological epochs, the higher forms preceded the lower. There was clearly room for degradation within a schema that was essentially progressive—a degradation which could be used to challenge any universal law of development.

4 If man had developed from more primitive life then either all creatures had an immortal soul or man did not. Neither option did Miller find palatable. Deprived of the prospect of eternal life, a belief in God would become irrelevant to one's earthly conduct.

5 Miller rejected this option because it conflicted with both the presuppositions and the spirit of the development hypothesis, the very origins of which were to be found in the renunciation of divine interference.

[1] F. N. Egerton, 'Refutation and Conjecture: Darwin's Response to Sedgwick's Attack on Chambers', *Studies in History and Philosophy of Science*, 1970, **1**, 176.

6 On one level the answer is yes, since he believed that the language of the Bible did not pretend to be scientific. Accordingly there could be no clash. On another level there clearly could be conflict if a 'scientific' theory had implications for the nature of man—implications which would undermine Biblical doctrine.

7 Since Powell considered the claims of science and the claims of theology to belong to separate domains of experience, he could see in Miller's work a last but spurious attempt at the harmonization of Genesis with Geology.

In the long run Powell's perspectives were to be the more viable, but in the short run Hugh Miller probably did more than any other nineteenth-century author to reassure his audience that geology had not obliterated the 'footprints of the creator'. In his book of that title, Miller could invoke the fossil Coelacanth which he had discovered (the asterolepsis of Stromness) to confound the exponents of development. They had presupposed, in common with the German idealists like Lorenz Oken, that 'no organism is, or ever has been created, which is not microscopic. Whatever is large has not been created but developed.' With the knowledge that his coelacanths were all of eight feet long, and with the evidence that they were the most ancient vertebrates that Scotland could boast, Miller's strategy was simple and compelling: 'In the not unimportant circumstance of size, the most ancient Coelacanths yet known, instead of taking their places, agreeably to the demands of the development hypothesis, among the sprats, sticklebacks and minnows of their class, took their place among its huge basking sharks, gigantic sturgeons, and bulky swordfishes. They were giants, not dwarfs.'

God the Creator had continued to create. But if Miller had lived to face the challenge of Darwin's theory, he would doubtless have shared the sentiment of his wife that 'the most efficient protest against this blind exclusive theory, which would inaugurate the reign of selfishness throughout nature, is to be found in the human heart.'[1]

In this section we have concentrated our attention on the responses of Sedgwick and Miller to the development hypothesis of Chambers. But there is one question which might still be at the back of your mind. How, for instance, would Sedgwick have described the independent acts of Creation which his own system required? How would he have reacted to the demand for such a description? Study text 24 in *Anthology I*, and then state your answer.

The answer appears to be that, for Sedgwick, the very demand is irreverent. He suggests that we can form no adequate conception of creative power. There are therefore intrinsic limits to scientific knowledge, and it is far better to acknowledge them than to mock the Creator with unsubstantial doctrines which had but spurious support from the 'dark corners' of nature.

---

6 The Balance of Nature

Almost all the great names you have encountered in this unit have at various times been hailed as precursors of Darwin. The very fact that they could diverge so much among themselves shows how shop-soiled the label is. But once the traditional line-up of precursors is disbanded—Buffon, Lamarck, Oken, Lyell, Chambers—new possibilities emerge. The rationale for the traditional line-up used to be a certain sympathy for

[1] H. Miller, *Footprints of the Creator*, sixteenth edition, Edinburgh, 1874. Prefatory remarks by Mrs Miller, xi.

'evolutionary' modes of thought, but also a supposed antipathy to conventional forms of natural theology. Those who indulged in evolutionary speculation were precursors of Darwin who could be arrayed *against* those natural theologians who had to bite the dust once Darwin's theory was approved. Those who proclaimed this message would have considered it self-contradictory to assert that a Christian natural theologian could have been a precursor. But this is one of the new possibilities that has to be examined once the traditional precursors are put in their place. Is it possible that Darwin's universe had closer affinites with the universe of the natural theologians than with that of anyone else? This is the question we must try to answer in this last section of the unit.

6.1 The natural theologians as predecessors

For many years Professor Hooykaas has argued that among those whom Gillispie maligns as 'catastrophists' were naturalists and geologists who contributed as much to the interpretation of the past as the so-called uniformitarians. Indeed it was the progressive creationists who at least admitted *some* progressive pattern among the fossils, which is more than can be said of Lyell, the champion of uniformity. But can one take the argument one step further? Is it possible to agree with Walter Cannon's provocative thesis that the true predecessors of Darwin were 'those scholars of the Anglican Church who had enough sense to worry about David Hume on miracles'?[1]

Put as baldly as that the thesis is obviously and deliberately over-stated. It also has the air of total implausibility! Hugh Miller was not an Anglican scholar, but he was a refined progressive creationist who had taken the trouble to refute David Hume (see Units 9–10). But he also took the trouble to refute schemes of development and transformation. Can it really make sense to align the Millers, Bucklands and Sedgwicks in the front rank of the 'forerunners'?

Cannon's revaluation of the bases of Darwin's achievement will not be easy for you to evaluate without a deeper knowledge of Darwin's theory, but we may be able to identify some of the strengths and weaknesses of his argument.

The triumph of Darwinism was, according to Cannon, none other than 'the triumph of a Christian way of picturing the world over the other ways available to scientists.' By this he does not mean that Darwin conceived his theory as an expression of Christian belief, but that the debate ignited by that theory was 'at bottom an argument as to whether or not to accept a secular version of a world described in Christian thought.'

Far from destroying the universe of the natural theologians, Darwin stole it from them: 'the subsequent wrath of the Christians is not the wrath of those who have been beaten by an alien; it is the wrath of those whose treasured possession has been stolen from them'. It was precisely because they had so much in *common* that the Darwinians and the theologians fought so fiercely: their conceptions of the universe were congruent, but where the natural theologians postulated God, Darwin proposed natural selection. It was not Cannon's intention to suggest that Christianity properly understood is compatible with Darwin's theory properly understood, but simply that from Christian natural theology Darwin inherited a set of ideas which defined a universe in which his mechanism of natural selection would as a matter of fact work. The true predecessors of Darwin were those who took their Paley seriously.

But how can this claim be substantiated? What was the crucial set of ideas? When Darwin said that he had memorized Paley at Cambridge but learned nothing from it does that refute or, in a more subtle way, confirm the thesis? Cannon himself claims to identify ten points of congruence between the universe of Darwin and that of the natural theologians. To analyse each one in turn would take us far beyond the scope of this unit, but if we select four of them, that should enable you to see the drift of the argument.

One feature which Darwin's universe is said to have in common with that of the natural theologians is the rejection of an anthropocentric perspective on the past. When

[1] W. F. Cannon, 'The Bases of Darwin's Achievement: A Revaluation', *Victorian Studies*, December 1961, 122.

Darwin argued that natural selection worked for the good of each animal, he was tacitly denying that all animals had appeared for the benefit of man. But this same denial, according to Cannon, had been made by every natural theologian worth his salt. You will recall from Units 9–10 that Boyle had indeed refrained from an extreme anthropocentrism, and among the progressive creationists of the nineteenth century it was commonly asserted that before man had stalked the earth there were countless creatures which had been adapted to enjoy life to their full. So the argument is that, unlike Lamarck who saw man as the goal of evolution, neither Darwin nor the natural theologians adopted man as the value-centre of creation.

The congruence manifests itself again in the context of purpose and adaptation. For Darwin the concepts of adaptation and adaptability were to be the touchstone of his theory: those creatures less well adapted to cope with the exigencies of their environment would perish in the struggle for survival. Accordingly those that did survive would of necessity be well adapted. In other words there was a sense in which Darwin *could* speak of natural selection's working for the good of every creature, as he did in early editions of the *Origin*: 'every detail of structure has been produced for the good of its possessor: . . . (any exception) would be absolutely fatal to my theory.'[1]

An organism might possess useless or rudimentary organs but they would not have been useless to its ancestors. The parallel with the universe of the natural theologians appears quite marked. It was naturalists of the calibre of John Ray who had pioneered the study of adaptation with the intention of documenting the wisdom of God. Though they interpreted it differently, the natural theologians had publicized purposive adaptation much as Darwin was bound to do.

A third respect in which parallels may be drawn is that concerning the direction and meaning of history. Darwin's reading of the fossil record was more profoundly historical than that of the speculative evolutionists who, like Lamarck, did not describe the process of evolution as essentially unrepeatable. For Darwin the fossil record, imperfect though it was, disclosed the details of a process which, starting from a single animate source, had occurred in time once and for all. The history of life was a unique history. In just such a way Christian theology had sponsored a profoundly historical interpretation of the universe—an interpretation with a sense of direction from Creation to Incarnation to Resurrection, through progressive Redemption to a final Consummation. This is a theme which Dr Richardson will develop in Unit 14, but for the moment it is sufficient to state the argument, presented both by Hooykaas and Cannon, that the historical dimension of the Christian natural theologians was a prerequisite of a Darwinian process which *locked* the history of life in the rocks.

Finally because the progressive creationism of the natural theologians was progressive, it was only necessary for Darwin to dilute and secularize any outstanding catastrophes in order to complete an explanation for the origin of species which could forsake all notion of *miraculous* creation. Cannon's argument is quite straightforward: 'It is of course the Catastrophist outline of the history of the earth and not that of Lyell which we more nearly accept today. Darwin did not have to convince his better scientific opponents of the existence of progressive development, for they had championed the idea for years.' Darwin did, however, set up an antithesis between natural selection and separate creation. Does this not vitiate the argument? Not in the least, Cannon affirms, since Darwin was misguided when he did so. For many of his predecessors the word 'creation' had come to mean nothing more than that unknown mechanism whereby living creatures had come into existence. That mechanism for a natural theologian must be of divine ordination, but that does not mean that Buckland, Sedgwick, and Whewell were campaigning for miracles.

That is the burden of the case for Darwin's debt to the orthodox natural theology of his day. The universe of Darwin, like that of Sedgwick, was one in which it made good sense to ask by what external organizing power living organisms had developed in close adaptation to their environments. Where Sedgwick saw selection divine, Darwin saw selection that need not be divine.

As a corrective to popular accounts of Darwin's 'precursors' Cannon's thesis is

[1] Cannon, *ibid.*, 124.

both engaging and stimulating. It cuts straight across the presuppositions of those historians whose sympathies lie with the warfare model. But just as extravagant claims have been made for the traditional precursors, so it is possible to overrate the insights of the natural theologians. The more closely one inspects the Darwinian mechanism the more incongruous it appears to be in the context of natural theology. You will have the opportunity in a moment to work this out for yourself, but it might be instructive if we first locate some of the objections that might be raised.

However congruent the universe of Darwin and of the natural theologians, they differed decisively on a point of order. If there were any irregularity in the history of life it was, for the natural theologians, an ordained irregularity. It was their most basic belief that the universe was ordered and therefore pre-ordained. But it was precisely this almost intuitive sense of order that was lacking in Darwin's vision of the past. Natural selection worked on *chance* variations in a manner which made the development of life a long tortuous and *random* affair. As G. G. Simpson has said of Darwin's theory, man became the product of a process that had never even had him in mind. This would seem to strike at the root of the congruence theory, and not least because it exposes an ambiguity in the word 'anthropocentric'. It is Cannon's contention that natural theologians like Buckland and Sedgwick were not anthropocentric in their outlook: they allowed that God's creatures were not fashioned for man's benefit alone, and they allowed patterns of creation in the past which did not all eventuate in man. But they would never have doubted that the past was a preparation for the present, that man had been created, in due season, after the image of God. The universe of the natural theologians was surely anthropocentric in the sense that man's uniqueness was still guaranteed by his place on the blueprint of creation? In Darwin's universe there may have been grandeur but there simply was no blueprint. For the same reason it is possible to exaggerate the supposed correlation between their concepts of purposive adaptation. When the natural theologians spoke of purposive adaptation they meant adaptation that had been purposed, designed: in Darwin's theory particular adaptations just happened to serve a sometimes transient purpose. These interpretations were worlds if not universes apart. It was of course possible to super-add a divine purposer to Darwin's scheme, but, as we shall see in Unit 13, this was not without its difficulties.

These are some of the objections that might be levelled at the congruence thesis. but the most telling is one you should be able to discover for yourself. As the starting point we ought to note that in his catalogue of *resemblances* between Darwin's universe and that of the natural theologians, Cannon includes this: 'the subservience of Malthusian superfecundity to the purposes of the system'. Once we have unravelled the meaning of this, you should be able to spot the problem.

6.2 The interpretation of Malthus

It is not surprising that Cannon should refer to Malthus in the context of evolutionary theory since historians have long seen connections between his *Essay on the Principle of Population* (1798) and the development of evolutionary theories. Nor is it simply that the *ideas* of Malthus might have influenced Darwin and Wallace. The essay sparked off lasting socio-political debates which, in England, contributed to the context for speculation about man's place in nature. It is quite impossible for us to do justice to the complexities of these repercussions, but we can perhaps see why there were some.

When the Reverend Thomas Robert Malthus wrote the first edition of his *Essay* it was to put constraints on the utopian speculations of Godwin and Condorcet who had promised the perfectibility of society, and to whose writings his own father had succumbed. No such theory could be taken seriously, Malthus argued, if it did not face up to the stark realities of life, one of which was the tendency for animal populations to increase more rapidly than their food supply. Bringing the authority of mathematics to his argument, Malthus proclaimed that food resources increasing arithmetically (1, 2, 3, 4 . . .) could not in the long run keep pace with a population increasing in geometric progression (1, 2, 4, 8, 16 . . .). With respect to man himself this meant

Figure 10 Caricature of Malthus: overpopulation, by George Cruikshank, from Comic Almanack, *1844–53.*

that preventive checks were desirable, such as the postponement of marriage and procreation—not by birth control but by abstinence. Otherwise it would be left to positive checks such as poverty, plague and famine to reduce a population to subsistence level. The articulation of the argument changed through successive editions, but the effect remained much the same: there was a lugubrious aspect to man's relationship with nature. In the context of natural theology the Malthusian argument could be interpreted as an indictment of a Providence which did not provide enough, but in the context of the Poor Laws the implications were even more contentious. Read the short extract below and see if you can say why. This is taken from chapter 5 of the first edition:

> The evil is perhaps gone too far to be remedied, but I feel little doubt in my own mind that if the poor laws had never existed, though there might have been a few more instances of very severe distress, yet that the aggregate mass of happiness among the common people would have been much greater than it is at present.
>
> Mr Pitt's Poor Bill has the appearance of being framed with benevolent intentions, and the clamour raised against it was in many respects ill directed, and unreasonable. But it must be confessed that it possesses in a high degree the great and radical defect of all systems of the kind, that of tending to increase population without increasing the means for its support, and thus to depress the condition of those that are not supported by parishes, and, consequently, to create more poor.

What political issue is raised by this passage?

The issue is the extent, if at all, to which the poor should be supported either by their parishes or by any special legislation. According to Malthus such support was misconceived because it encouraged population increase without augmenting the means of subsistence.

Figure 11 Thomas Malthus, portrait by J. Linnell (Mansell Collection).

'Were I to propose a palliative,' Malthus continued, 'and palliatives are all that the nature of the case will admit, it should be, in the first place, the total abolition of the present parish-laws.' Without going into details it should be clear by now why the name of Malthus was so frequently summoned when the distress of the poor was reviewed. Pitt did drop his bill for extending poor relief to large families.[1] Rich and poor alike were part of nature, but it was the poor who would inevitably suffer: not a message that was likely to appeal to social reformists. Without mincing his words, William Cobbett wrote to Malthus: 'Parson, I have, during my life, detested many men: but never any one so much as you.' Sensitive to the wretched conditions of the English working class Engels could see in the Malthusian law of population 'the most open declaration of war of the bourgeoisie upon the proletariat. . . . If then the problem is not to make the "surplus population" useful, to transform it into available population, but merely to let it starve to death in the least objectionable way and to prevent its having too many children, this, of course, is simple enough, provided the surplus population perceives its own superfluousness and takes kindly to starvation.'[2]

But what bearing does this have on the question of Darwin's debt to natural theology? What does Cannon mean by 'the subservience of Malthusian superfecundity to the purposes of the system'? Each of the following extracts provides an example

[1] R. M. Young, 'Malthus and the Evolutionists: the Common Context of Biological and Social Theory', *Past and Present*, 1969, **43**, 109–45.

[2] R. M. Young, 'The Historiographic and Ideological Context of the Nineteenth-Century Debate on Man's Place in Nature', in M. Teich and R. M. Young (eds.), *Changing Perspectives in the History of Science*, Heinemann, 1973, p. 382.

of what he probably means. Study each one in turn, noting the author of each. Then answer this question: would you say that the way in which Darwin interprets Malthus, aligns him with or separates him from the natural theologians?

To urge man to further the gracious designs of Providence by the full cultivation of the earth, it has been ordained that population should increase much faster than food. This general law . . . undoubtedly produces much partial evil, but a little reflection may, perhaps, satisfy us, that it produces a great overbalance of good. Strong excitements seem necessary to create exertion, and to direct this exertion, and from the reasoning faculty, it seems absolutely necessary, that the Supreme Being should act always according to general laws . . . if we return to the principle of population and consider man as he really is, inert, sluggish, and averse from labour, unless compelled by necessity , . . we may pronounce with certainty that the world would not have been peopled, but for the superiority of the power of population to the means of subsistence. Strong and constantly operative as this stimulus is on man to urge him to the cultivation of the earth, if we still see that cultivation proceeds very slowly, we may fairly conclude that a less stimulus would have been insufficient. . . . Had population and food increased in the same ratio, it is probable that man might never have emerged from the savage state. But supposing the earth once well peopled, an Alexander, a Julius Caesar, a Tamberlane, or a bloody revolution might irrecoverably thin the human race, and defeat the great designs of the Creator. (*Malthus*)

But then this *superfecundity*, though of great occasional use and importance, exceeds the ordinary capacity of nature to receive or support its progeny. All superabundance supposes destruction or must destroy itself. Perhaps there is no species of terrestrial animals whatever which would not overrun the earth, if permitted to multiply in perfect safety; or of fish, which would not fill the ocean—at least, if any single species were left to their natural increase without disturbance or restraint, the food of other species would be exhausted by their maintenance. It is necessary therefore, that the effects of such prolific faculties should be curtailed. . . . Though there may be the appearance of failure in some of the details of nature's works, in her great purposes there never are. (*Paley*)

In October 1838, that is, fifteen months after I had begun my systematic enquiry, I happened to read for amusement Malthus on *Population*, and being well prepared to appreciate the struggle for existence which everywhere goes on from long-continued observation of the habits of animals and plants, it at once struck me that under these circumstances favourable variations would tend to be preserved, and unfavourable ones to be destroyed. The result of this would be the formation of a new species. Here, then, I had at last got a theory by which to work: but I was so anxious to avoid prejudice, that I determined not for some time to write even the briefest sketch of it.

De Candolle, in an eloquent passage, has declared that all nature is at war, one organism with another, or with external nature. Seeing the contented face of nature, this may at first well be doubted; but reflection will inevitably prove it to be true. The war, however, is not constant, but recurrent in a slight degree at short periods, and more severely at occasional more distant periods; and hence its effects are easily overlooked. It is the doctrine of Malthus applied in most cases with tenfold force. As in every climate there are seasons, for each of its inhabitants, of greater and less abundance, so all annually breed; and the moral restraint which in some small degree checks the increase of mankind is entirely lost.

That there is much suffering in the world no one disputes. Some have attempted to explain this with reference to man by imagining that it serves for his moral improvement. But the number of men in the world is as nothing compared with that of all other sentient beings, and they often suffer greatly without any moral improvement. This very old argument from the existence of suffering against the existence of an intelligent First Cause seems to me a strong one; whereas, as just remarked, the presence of much suffering agrees well with the view that all organic beings have been developed through variation and natural selection. (*Darwin*)

What is your answer to the question above, and how would you justify it?

There is undoubtedly a superficial resemblance between Malthus, Paley and Darwin in that they each exploit the population principle for their own ends, embracing it within a wider 'system' of thought. But that is as far as the congruence goes. Both Malthus and Paley *absorb* the problem of struggle within their respective systems of natural theology, whereas Darwin *magnifies* it: nature is the scene of recurrent war: 'it is the doctrine of Malthus applied in most cases with ten-fold force.' For Paley there had to be checks to overpopulation simply to maintain the *balance* of nature: Malthus too claims that his law is in perfect harmony with the gracious designs of Providence, producing as it does a great 'overbalance of good'. In Darwin's discussion, however, there is no mention of balance: the 'contented face of nature' is but a mask. Favourable variations would be preserved, unfavourable ones destroyed—that was all that could be said. The contrast solidifies in Darwin's discussion of suffering. Malthus had himself urged moral improvement as the justification for his law: man was naturally sluggish and required incentives to work. But, as Darwin insisted, this argument had no force when applied to other animals. For Paley the *appearance* of suffering had to be explained away: for Darwin the *reality* of suffering was already explained.

These considerations would seem to go a long way towards undermining the congruence thesis that the true predecessors of Darwin were the natural theologians who took their cue from Paley. Whereas they presupposed the balance of nature, Darwin presupposed an imbalance—a nature which 'may be compared to a surface on which rest ten thousand sharp wedges touching each other and driven inwards by incessant blows.'

This is not to say that Darwin was indebted to no one. He was a man of his time as we all are of ours. As Robert Young has stressed, Darwin did take Paley's answers; but he converted them into questions. The story of that conversion properly belongs to your next unit.

Further Reading

W. Coleman, *Biology in the Nineteenth Century*, Wiley, New York, 1971.
E. Gasking, *The Rise of Experimental Biology*, Random House, 1970.
J. C. Greene, *The Death of Adam*, Mentor, New York, 1961.
A. O. Lovejoy, *The Great Chain of Being*, Harper, New York, 1960.
H. L. McKinney, *Lamarck to Darwin*, Coronado Press, Kansas, 1971.
M. J. S. Rudwick, *The Meaning of Fossils*, Macdonald, 1972.

Acknowledgements

It is a pleasure to acknowledge both the advice of members of the course team, and my debt to Michael Bartholomew of the University of Lancaster for numerous constructive suggestions.

Grateful acknowledgement is made to the following sources for material used in this unit:

Text

Excerpts from *Darwin's Century* by Loren C. Eiseley. Copyright © 1958 by Loren C. Eiseley. Reprinted by permission of Doubleday & Company, Inc and Victor Gollancz Ltd.

Illustrations

Figure 1: National Portrait Gallery; *Figures 2, 3, 4, 5, 6, 7 and 11:* Mansell Collection; *Figure 8:* Radio Times Hulton Picture Library.

Unit 13
Darwin

Prepared by John Hedley Brooke for the Course Team

ILLUSTRATION OVERLEAF: *Charles Darwin, watercolour by George Richmond, 1840 (Down House Collection. By kind permission of the President and Council of the Royal College of Surgeons of England).*

Contents

1 Introduction and Objectives

The name of Darwin has often been associated with those of Marx and Freud, to indicate his seminal influence on man's understanding of himself. As the naturalist who made the idea of man's evolution intellectually respectable, he has invariably been praised as one of the great heroes of science. It has also become a cliché to assert that Darwin's theory ran through Victorian society like a plough through an anthill: everyone came running forth, so A. D. White reported, 'angry and confused'. Not all historians have agreed, however. The late Charles Raven referred to the Darwinian debate as no more than a storm in a Victorian tea cup. Consequently the main object of this unit is to help you to come to your own conclusions about the conception, originality and impact of Darwin's work. What was the structure of his theory, and how was his *On the Origin of Species* received when the first edition appeared in 1859?

The details and ramifications of Darwin's achievement are so intricate and complex that we cannot possibly do them justice in one brief unit. You will undoubtedly have questions in your mind which will be left unanswered when you reach the last page, and it will be all to the good if you have more than when you started! Units 15 and 16 will take you further; we can, however, select a cross-section of problems which should give you some insight into the development of Darwin's theory, its strengths and weaknesses, and the reasons for its notoriety and success. The object is not to construct a comprehensive biography of Darwin, but to introduce you to some of the crucial issues which the historian has to consider.

We shall of course make passing reference to Darwin's education, and this will be the object of Section 2. His father, Robert Waring Darwin, an imposing man of six feet and two inches and weighing all of twenty-four stones, had established a flourishing medical practice at Shrewsbury and naturally encouraged Charles to follow his elder brother to the medical school of Edinburgh University. One of his teachers there was Robert Jameson (see television programme 7). Disenchanted with the life of a medical student, Charles was to leave Edinburgh for Cambridge where, in conformity with his father's wishes, he was prepared to study for an alternative career that would befit a gentleman: the Anglican ministry. On graduating from Cambridge, however, his father's intentions were foiled once again when Charles was recommended for the post of naturalist on HMS *Beagle*. Under the command of Robert FitzRoy, the *Beagle* was about to sail for South America where an exercise in surveying had to be completed.

Figure 1 Dr Robert Darwin, from a contemporary silhouette c. *1826* (Autobiography of Charles Darwin, *edited by Nora Barlow. Courtesy of William Collins Ltd.*).

For Darwin it was to be a voyage of great discovery, and in Section 3 you will have an opportunity to consider the observations and experiences which had the most profound influence upon him, preparing the way for a theory of speciation which began to take shape a year or so after his return home. The vessel was away from England for almost five years in all, and during that time, Darwin had been wrestling with all the problems posed by the geographical distribution of species. Why were different but indisputably related forms to be found in adjacent areas; what was the significance of the resemblance between fossil and living species in one and the same area; why, in the Galapagos Archipelago, were there flora and fauna peculiar to each island? Preoccupied with questions of this kind, and prompted by the geographical writings of Charles Lyell, Darwin eventually saw that a theory of descent with modification would explain his data. His first notebook on the transformation of species was opened in 1837, and in Section 4 you will be able to examine his state of mind at that time. The mainspring of the theory was to be the natural selection of favourable variations in the struggle for existence—a thesis which owed something to his reading of Malthus (cf. Unit 12) in the autumn of 1838. Just what that something was you will be able to consider for yourself.

Darwin's mechanism for evolutionary change was an imaginative synthesis: it presupposed both a struggle for existence and the occurrence of favourable variations—variations which might be preserved at the expense of normal or inferior members of a given population. Consequently, for the insemination of his theory Darwin had to gather as much information concerning patterns of variation as he could lay his hands on. You may be surprised to find him consorting with gardeners and pigeon-fanciers, but in Section 5 you will see why this was: if domestic breeders could accumulate small variations by their methods of selection, could nature not do likewise? Darwin's choice of the metaphor 'natural selection' testifies to his adoption of an analogy between man's selection of varieties under domestication, and nature's selection in the wild. Comparing his theory with that of Alfred Russel Wallace, who almost beat him to the post, Darwin ventured to say that his own theory had actually *originated* with his study of variation under domestication. There were, however, several problems with the analogy, and in Section 6 you will be encouraged to think about them.

These problems were made all the more difficult as a consequence of an underlying problem in the philosophy of science. The role of metaphors and analogies in the construction of scientific theories is a perennial subject of debate among philosophers. Do such analogies serve only a heuristic or illustrative function? Are they merely expository devices? Is their *scientific* function perhaps restricted to the *suggesting* of hypotheses? Or can they be a constitutive part of the theory itself? And, for that matter, can a theory ever be anything more than a set of analogies—devised to explain the unfamiliar by the familiar? Since one's attitude towards the status of Darwin's theory could depend on one's answers to these questions, it is important to investigate how Darwin himself uses the analogy. At the heart of this unit is Section 7 which is concerned with Darwin's argument as he summarized it in the last chapter of the *Origin*. You will be invited to study his exposition of the merits of the theory, and to consider the scientific objections that were levelled against it. Between the publication of the *Origin* and the *Descent of Man* (1871), Darwin was to readjust his mechanism, revoking his former emphasis on the primacy of natural selection and attaching an increasing importance to the direct action of environmental conditions in inducing responses of a Lamarckian kind (cf. Unit 12). The evidence suggests, however, that Darwin modified his theory off his own bat, rather than in response to criticism. Of course, not all objections were ostensibly scientific, and in Section 8 we shall examine some of a more philosophical nature. The phrase which Darwin borrowed from Herbert Spencer—'the survival of the fittest'—has often been attacked for its circular logic. To say that the fittest survive in the struggle for existence is, so it has been alleged, to say nothing of note since what we mean by fitness is the ability to survive! Whether this was, and is, a valid objection, you should by then be in a position to decide for yourself.

Opposition from the general public was likely to be less refined, and popular religious sentiments were for the most part hostile. Accordingly, in Section 9, you will have the chance to write your own review of Darwin's work—or two to be precise: one from the point of view of an enthusiastic naturalist; another from the standpoint of an angry cleric. On a more elevated plane the problem that Darwin's theory raised for theology was not whether there was a Mind behind the evolutionary process, but what sort of Mind it could be. Did it still make sense to claim evidence of beneficent design, when Darwin's portrait of the past was of nature *groping* her way forward? Or as one contemporary explained it: 'if bad only becomes good by steady starvation of the worse, (is it) possible to see in this process the cherishing love of a divine Creator?'[1] This was of course a caricature, but there is no denying the problem. In Section 10 you will be asked to evaluate different responses to this very problem: could natural selection be reconciled with natural theology?

Despite the religious objections, and in part because of some religious approval, Darwin's theory very rapidly won acclaim. Its successful dissemination was a consequence of many circumstances: its scientific pedigree and explanatory power, its controversial application to man himself and the ease with which it could be assimilated by often divergent social and political ideologies. These are the points which we shall develop in Section 11 where you should be able to detect some of the ulterior purposes for which the name of Darwin was invoked.

[1] R. H. Hutton, *Aspects of Religious and Scientific Thought*, London, 1899. Gregg International Reprint, 1971, p. 48.

With that structure of the unit in mind we can now spell out our main objectives. After completing this study you should be able to:

1 Discuss the significance of Darwin's voyage for the development of his theory, and explain why he was amazed by what he found on the Galapagos Islands;

2 Discuss the state of Darwin's thinking in 1837 as he filled his first notebook on transmutation, and explain how the reading of Malthus could have been relevant to Darwin's search for an evolutionary mechanism;

3 State some of the problems which Darwin had to solve in the development of his theory, and explain why he took such a keen interest in the study of variation under domestication;

4 Analyse the problems associated with Darwin's analogy between human selection and natural selection;

5 State the main lines of evidence which convinced both Darwin and Wallace that evolution by natural selection had in fact occurred, and tabulate the facts which Darwin considered to be explained better by his theory than by a theory of separate creation;

6 Give examples of the most serious scientific and philosophical objections that were levelled at Darwin's thesis;

7 Explain why there was a conflict between the Darwinian theory and *popular* religious sentiments, and evaluate arguments which purported to show that natural selection was perfectly compatible with natural theology;

8 Account for the success of Darwin's theory, with particular reference to the social and political ideologies with which it could become associated.

2 A Renegade Scholar?

'It seems . . . that a taste for collecting beetles is some indication of future success in life!' In an autobiographical sketch Darwin claimed that he had been born a naturalist. At his first school in Shrewsbury he was already collecting minerals, newts and pebbles; when he completed his formal education at Cambridge 'no pursuit was followed with nearly as much eagerness or gave me so much pleasure as collecting beetles'. An independent mind, a keen eye, a passion for collecting and classifying, and a resistance to formal study are characteristics of Darwin's youth which all his biographers have detected. At school, having read Gilbert White's *Natural History of Selborne* (1789), he wondered why every gentleman did not become an ornithologist. The fact that at Edinburgh he preferred shooting to medicine, and at Cambridge beetles, botany and dissipation to a dedicated preparation for the ministry, has made him a classic case of a ne'er-do-well whose subsequent achievements far outstripped his promise as a scholar. Nor is this picture seriously inaccurate. He could say of his three years at Cambridge that 'my time was wasted, as far as the academical studies were concerned, as completely as at Edinburgh and at school'. Although his father may have reacted with severity to the prospect of an idle sportsman for a son, Charles did confess that during his time at Edinburgh 'I became convinced from various small circumstances that my father would leave me property enough to subsist on with some comfort', and this belief 'was sufficient to check any strenuous effort to learn medicine'.

But it is possible to overplay the image of the renegade scholar. He may have been bored by his lecturers but that is scarcely unusual. Besides, 'to my mind there are no advantages and many disadvantages in lectures compared with reading'. When forced into contemplating a career in the Anglican ministry, 'I read with care "Pearson on the Creeds", and a few other books on divinity; and as I did not then in the least doubt the strict and literal truth of every word in the Bible, I soon persuaded myself that our Creed must be fully accepted.' When proceeding to Cambridge he could translate Homer and the Greek Testament, and when preparing for his final degree he derived as much pleasure from the logic of Euclid as from the logic of Paley. It is true that his celebrated walks with Henslow, the resident Professor of Botany, taught him more of entomology, mineralogy and geology than could the printed word; at the same time what books he did read could make a profound impression upon him. He later confessed that Linnaeus and Cuvier had been his gods; and 'during my last year at Cambridge, I read with care and profound interest Humboldt's *Personal Narrative*. This work, and Sir J. Herschel's *Introduction to the Study of Natural Science*, stirred up in me a burning zeal to add even the most humble contribution to the noble structure of Natural Science'. His imagination was fired by Humboldt's travels, and he was soon making plans with his cousin, William Darwin Fox, to sail to Tenerife in June 1832. However, as Darwin himself put it, 'this scheme was, of course, knocked on the head by the voyage of the *Beagle*'.

Rarely has a burning zeal found so perfect an outlet! Darwin could hardly believe his good fortune when, after a geological tour with Adam Sedgwick, he received an invitation which would take him further than the mountains of North Wales, far further even than the Canary Islands. Would he join HMS *Beagle* as the ship's naturalist? That Henslow had recommended him so warmly is a further reminder that Darwin had more than made his mark in Cambridge. Having been permitted to associate with men 'so much older than me and higher in academical position', and here Darwin had Whewell in mind as well as Sedgwick and Henslow, 'I infer that there must have been something in me a little superior to the common run of youths'. What that something was still escaped his father who was at first convinced that a voyage round the world was a wild scheme that would divert his son's energies yet again. Not only that but such a venture would be disreputable to the character of a prospective clergyman. With due deference to his father's wishes Charles had to decline the invitation. But all was not yet lost. His uncle, Josiah Wedgwood, came to Darwin's aid, answering all the objections that could possibly be mustered. To 'my dear doctor', he wrote: 'I should not think it would be in any degree disreputable to his character as a clergyman. I should on the contrary think the offer honourable to him; and the pursuit of Natural History, though certainly not professional, is very suitable to a clergyman.' The opportunity was not to be missed: the undertaking might be useless as regards his profession, 'but looking upon him as a man of enlarged curiosity, it affords him such an opportunity of seeing men and things as happens to few'. Little did his uncle know, when he wrote those words, that as a consequence of the voyage Darwin would see men and things as no man had before. His father having relented, and with the knowledge that £500 ought to see him through the whole voyage, Darwin joined Captain FitzRoy whom he found 'everything that is delightful'.

3 The Voyage of Discovery

It is one of those curious ironies of history that when, on 27 December 1831, the *Beagle* finally set sail, both her captain and her naturalist were agreed that whatever scientific discoveries they were likely to make would harmonize with, and in FitzRoy's case even substantiate the Bible.

How different from Darwin's later perspective: 'Considering how fiercely I have been attacked by the orthodox, it seems ludicrous that I once intended to be a clergyman. Nor was this intention and my wish ever formally given up, but died a natural death when . . . I joined the *Beagle* as naturalist.'

It is quite impossible in a few lines to recapture the exhilaration, the awe, the

Figure 2 The Beagle, *with Mount Sarmiento, Tierra del Fuego, in the background, 1838 (Mansell Collection).*

spirit of adventure, and the changing vistas that were Darwin's life for the next five years. Among the scenes that remained deeply impressed on his mind 'none exceed in sublimity the primeval forests undefaced by the hand of man; whether those of Brazil, where the powers of life are predominant, or those of Tierra del Fuego, where Death and Decay prevail. Both are temples filled with the varied productions of the God of Nature: no-one can stand in these solitudes unmoved, and not feel that there is more in man than the mere breath of his body.' There really is no substitute for reading Darwin's own account of *The Voyage of the Beagle.* With the privilege of hindsight one can of course speculate about the way in which different experiences might have informed his later theory, though one must be constantly on guard against reading the story backwards. It was during the voyage that he came to realize that there was nothing like geology. 'The pleasure of the first day's partridge shooting or the first day's hunting cannot be compared to finding a fine group of fossil bones, which tell their story of former times with almost a living tongue. . .'. For the interpretation of that tongue Lyell was his guide. Volume 1 of the *Principles of Geology* went with him from the start, and volume 2 was despatched to him at Montevideo. What did Darwin himself find? At Punta Alta he was 'very successful with the bones', locating parts of the giant sloth and two equally large monsters. There were remains of giant armadillos and of wild llamas the size of camels. How could Darwin not be

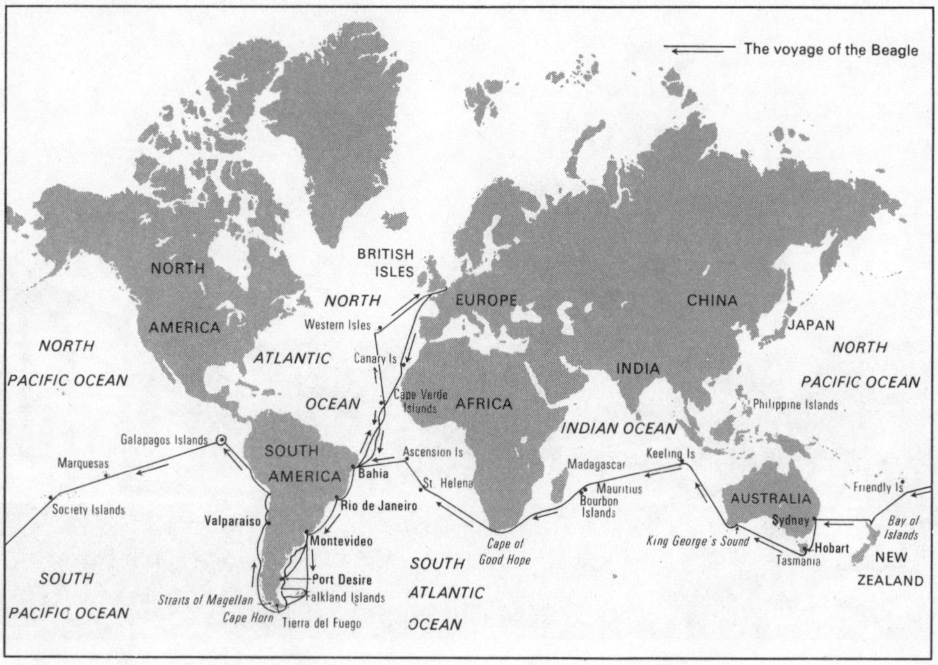

Map tracing the voyage of the Beagle.

impressed by these gargantuan remains which so closely resembled the smaller sloths, armadillos and guanacos which surrounded him? 'This wonderful relationship in the same continent between the dead and the living will', he later declared, 'hereafter throw more light on the appearance of organic beings on earth and their disappearance from it, than any other class of facts.' He could not conceal his 'deepest astonishment.' Indeed, 'if Buffon had known of the gigantic sloth and armadillo-like animals... he might have said with a greater semblance of truth that the creative force in America had lost its power, rather than that it had never possessed great vigour'.

The affinities between the living and the dead had struck Darwin as they had Lamarck over thirty years before. But with this difference, that Darwin's dead were more than dead—they were indisputably extinct. Nor was it the mere fact of extinction that excited his imagination. The astonishing reality was the extermination of so *many* species. 'No fact in the long history of the world', he was to write, 'is so startling as the wide and repeated extermination of its inhabitants.' Drawing on Lyell it was not difficult for Darwin to explain how extinction could occur as the result of quite inappreciable causes; but, as Jonathan Hodge has indicated, there was one respect in which Lyell failed him. How could one explain the striking fact, disclosed by fossil remains in a particular area on the coast of Argentina, that certain mammals had arisen and become extinct *within* a larger period of time defined by the careers of certain mollusc species in that same area? That the molluscs were still extant, together with the fact that other living mammals, such as the armadillos, bore so close a resemblance to their allied fossil species, suggested that no substantial change in conditions had occurred. So why had *some* mammals perished altogether? Was their non-adaptation to new conditions a realistic explanation? Darwin's response to this problem is discussed by Dr Hodge in television programme 10. For a while he was to consider the idea that new species might be conceived as grafts on old ones: as the life-span of fruit and vegetable grafts was limited, so each natural species might have its own life-span impressed upon it—an analogy which might have conditioned Darwin's first speculations about the possibility of common descent. Certainly he was not yet applying the doctrine of Malthus with ten-fold force. Not until he was back in England did he read Malthus for himself. We should not, however, ignore his own insistence that he was well *prepared* to appreciate the struggle for existence. It has often been suggested that he had been well prepared by the very fact of living in England at a time when ruthless exploitation was rife, and when the most squalid aspects of industrialization were all too clear. This may well be so, but one cannot help thinking that his voyage around the world was a better preparation still. His *Origin of Species* was to be replete with the images of warfare. Was it not in the

Figure 3 Cathedral at Concepcion after the earthquake, by J. C. Wickham, from Narrative of the Surveying Voyages of HMS Adventure and Beagle *(from Alan Moorehead,* Darwin and the Beagle. *Courtesy of George Rainbird Ltd.).*

Pampas that he had witnessed the war between land-hungry Argentinians and the native Indians whose plight was intensified by the massacres of General Rosas? Was it not at the very tip of South America that Darwin had witnessed the native Fuegians battling with the bleakest elements? It was a consequence of the voyage that Darwin saw nature in the raw, saw the puma devour the guanaco, the giant condor preying on sheep and cattle. Several of his journeys on foot were at the risk of being shot, or savaged by tigers, and as if this were not enough to establish the malign indifference of nature, he actually saw an earthquake—the most violent that could be remembered in the vicinity of Concepcion.

Was it possible to believe in a stable and a divinely-balanced nature, when the cathedral at Concepcion lay smashed in ruins? Darwin was to associate this scene with thoughts of home: 'If beneath England the now inert subterranean forces should exert those powers, which most assuredly in former geological ages they have exerted, how completely would the entire condition of the country be changed! . . . In every large town famine would go forth, pestilence and death following in its train.'

While we are still speculating we ought to take heed of what Darwin himself regarded as 'the most curious and interesting spectacle I ever beheld'. This was the sight of native Fuegians gesticulating on the shore line. 'I could not have believed how wide was the difference between savage and civilized man; it is greater than between a wild and a domestic animal, inasmuch as in man there is a greater power of improvement.' The reference to a power of improvement is of particular interest since on board ship there had been three Fuegians who had been 'civilized' in England and who were now being returned by Captain FitzRoy to their homeland.

Figure 4 Fuegian types studied and drawn by FitzRoy in the appendix to the Narrative of the voyage (from Alan Moorehead, Darwin and the Beagle. *Courtesy of George Rainbird Ltd.).*

The contrast between the dandy Jemmy Button and the cannibals who were his brothers made a deep impression on Darwin, and might perhaps have made it easier for him to strip the veneer from civilized man and lay bare the civilized animal—the very step that Lyell was not prepared to take. When reflecting on the pleasures and pains of his circumnavigation, Darwin returned to the same point:

> Of individual objects, perhaps nothing is more certain to create astonishment than the first sight in his native haunt of a barbarian. . . . One's mind hurries back over past centuries, and then asks, Could our progenitors have been men like these?—men, whose very signs and expressions are less intelligible to us than those of the domesticated animals.

We know what answer Darwin gave to this question, but not before he had conceived a mechanism for the derivation of one species from another. It was not until July 1837 that Darwin opened his first notebook on transmutation, but the idea of fixed, immutable species was one which he was almost forced to question after his visit to the Galapagos Islands.

Of the natural history of these islands Darwin wrote:

> Most of the organic productions are aboriginal creations, found nowhere else; there is even a difference between the inhabitants of the different islands; yet all show a marked relationship with those of America, though separated from that continent by an open space of ocean between 500 and 600 miles in width. . . . Considering the small size of these islands, we feel the more astonished at the number of their aboriginal beings, and at their confined range. Seeing every height crowned with its crater, and the boundaries of most of the lava streams still distinct, we are led to believe that within a period geologically recent the unbroken ocean was here spread out. Hence, both in space and time, we seem to be brought somewhat near to that great fact—that mystery of mysteries—the first appearance of new beings on this earth.

Could Darwin get nearer still? Never having dreamed that islands in such close proximity, and exhibiting the same geographical features, could have been 'differently tenanted', he expressed his amazement at what he found:

> The distribution of the tenants of this archipelago would not be nearly so wonderful if, for instance, one island had a mocking-thrush, and a second island some other quite distinct genus: if one island had its genus of lizard, and a second island another distinct genus, or none whatever; or if the different islands were inhabited, not by representative species of the same genera of plants but by totally different genera. . . . But it is the circumstance that several of the islands possess their own species of the tortoise, mocking-thrush, finches, and numerous plants . . . that strikes me with wonder.

Why do you think Darwin could express himself in this way? Why does he find the actual distribution more wonderful than the alternatives he mentions?

The remarkable point was that the action of 'creative force'—as Darwin provisionally called it—could be so diverse and yet analogous on points so near each other. This was more wonderful than either complete diversity or complete identity, not only because it was so unexpected but perhaps because it raised this question: why had one Creator fashioned different forms to fill essentially identical niches in the economy of nature?

It is possible that Darwin might have interpreted these facts in terms of God's ingenious care, but he was to become increasingly convinced that if there was rhyme and reason in the universe, the distribution of species ought to be different from what it actually was. Against Lyell's proposition that species were created when and where the conditions were right, Darwin was to parade instances of perfect conditions going to waste—where the appropriate life was conspicuous by its absence. Returning to the Galapagos: had one species been taken and *modified* for different ends on different islands? This was to be Darwin's eventual speculation after studying the Galapagos finches and mocking-birds. In the different species of *Geospiza* he found a perfect gradation in the size of beaks, from one as large as that of the hawfinch to that of a chaffinch and even to that of a warbler. Here was information that he would later use to advantage: 'Seeing the graduation and diversity of structure in one small, intimately related group of birds, one might really fancy that, from an original paucity of birds in this archipelago, one species had been taken and modified for different ends.'[1]

There is little doubt that here was one starting point for Darwin's deliberations on the transmutation of species, but we must be wary of oversimplifying the logic of discovery. To say that Darwin proceeded to devise a theory which would explain his collected facts is only a half-truth. The mistake in such a view is to suppose that the facts are somehow 'given'. Darwin himself knew better than that. One of the things that had worried him earlier in the voyage was whether he was noting the right facts. Whether a fact is interesting or relevant depends on one's theoretical framework, and nowhere is this better illustrated than by Darwin's collection of the Galapagos data.

Figure 5 Darwin's finches: an illustration of natural selection. The fourteen species of Darwin's finches are arranged to suggest the evolutionary tree of their development, divided broadly into ground finches (geospiza), *the earliest form, and tree finches. Of the tree species, 1 is woodpecker-like, 2 inhabits mangrove swamps, 3, 4 and 5 are insect-eating, 6 is vegetarian, 7 is a single species of warbler-finch, 8 is an isolated species of Cocos Island finch. Of the ground finches, 9, 10 and 11 are seed-eaters, 12 is sharp-beaked, 13 and 14 are cactus-eaters (from* Scientific American, *April 1953. Courtesy of David Lack and* Scientific American).

[1] *Journal of Researches*, 1845, 380.

The truth is, as he acknowledged himself, that until it was almost too late he virtually disregarded a remark made by the Vice-Governor, Mr Lawson, to the effect that he could tell merely by looking at a tortoise from which island it had come. Consequently, Darwin had partially mingled his collections before the full significance of that remark sank in. And ironically this was true of his data on the finches. 'Unfortunately most of the specimens of the finch tribe were mingled together; but I have strong reasons to supect that some of the species of the sub-group *Geospiza* are confined to separate islands.' Darwin was left to suspect.

4 The Beginnings of a Theory

After arriving home in October 1836 it took Darwin a full ten years to describe and classify his *Beagle* materials, and a further eight were to be devoted to the study and classification of barnacles. His doubts concerning the stability of species were, however, within the space of two years transformed into a working theory of descent with modification (see television programme 10). That the transformation had not taken place earlier—indeed on the voyage itself—is partly due to the fact that Darwin had not yet found an explanation for the perfect adaptations he had seen everywhere about him. Finches might differ from island to island, but how had each species been adapted to its own ecological niche? With this question before him, Darwin opened his first notebook on the transmutation of species in July 1837. Earlier that year he had reread Lyell's *Principles of Geology*, now in its fifth edition, making his own critical annotations and referring for the first time to 'my theory'. For an insight into Darwin's frame of mind as he filled his notebook, you might like to peruse the following excerpts. The word 'propagation' in the first extract probably meant some kind of common descent—by analogy with budding and grafting on one stock (see television programme 10).

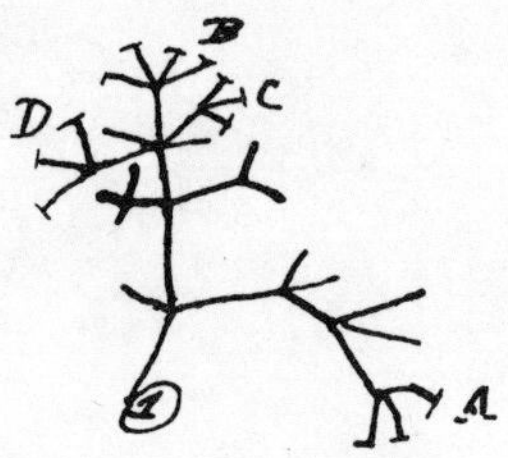

Figure 6 First diagram of the tree of evolution, from First Notebook on Transmutation of Species, *1837 (courtesy of the Trustees of the British Museum (Natural History)).*

> Propagation explains why modern animals same type as extinct, which is law, almost proved.
>
> If we choose to let conjecture run wild, then animals—our fellow brethren in pain, disease, death, suffering and famine, our slaves in the most laborious works, our companions in our amusements—they may partake from our origin in one common ancestor, we may be all netted together.
>
> There never may have been grade between pig and tapir, yet from some common progenitor. Now if the intermediate ranks had produced infinite species, probably the series would have been more perfect.
>
> Cuvier objects to propagation of species by saying, why have not some intermediate forms been discovered between Palaeotherium, Megalonyx, Mastodon, and the species now living? Now according to my view [in S. America] parent of all Armadillos might be brother to Megatherium—uncle now dead.
>
> Opponents will say—show them me [intermediate forms]. I will answer yes, if you will show me every step between bulldog and greyhound.
>
> All animals of same species are bound together just like buds of plants, which die at one time, though produced either sooner or later. Prove animals like plants—trace gradation between associated and non-associated animals—and the story will be complete.
>
> With respect to extinction, we can easily see that (a) variety of [the] ostrich [Petise], may not be well adapted, and thus perish out; or, on the other hand, like Orpheus [a Galapagos bird], being favourable, many might be produced. This requires [the] principle that the permanent variations produced by confined breeding and changing circumstances are continued and produce(d) according to the adaptation of such circumstances, and therefore that death of species is a consequence (contrary to what would appear from America) of non-adaptation of circumstances.

> When one sees nipple on man's breast, one does not say some use, but sex not having been determined—so with useless wings under elytra of beetles—born from beetles with wings, and modified—if simple creation merely, would have been born without them.[1]

What can you infer from these extracts about the state of Darwin's thinking at this time?

He had evidently conceived the possibility that man and the animals might all have descended from one common ancestor—an idea that would explain 'why modern animals same type as extinct'. It was already clear to him that a resemblance between two living creatures (e.g. tapir and pig) did not necessarily mean that the one had descended from the other: it was possible they might have diverged from a common progenitor. He was in a position to challenge Cuvier's objection concerning the absence of intermediate forms—an objection which he could thwart by reference to domestic species. He appears to be working on the assumption that a certain 'term of life' had been impressed on each species, and he has conceived what would be required to complete the theory. There is a hint of a return to the idea that a species would become extinct if it was no longer adapted to its circumstances, and the adoption of the principle that variations could become permanent within a confined population. Finally there is Darwin's suggestion that the existence of useless organs was more in keeping with his own theory than with that of 'simple creation'.

In all this there is of course a distinct break from any notion of a 'chain of being' (cf. Unit 12). Since Darwin also uses the traditional language of creation at this time, one cannot read the *Origin* out of the notebook, but two points at least stand out from these extracts: Darwin's concern with the problem of variation and its permanence; and his almost melancholic preoccupation with 'pain, disease, suffering and famine'. This would certainly seem to corroborate his own testimony that when he read Malthus in the autumn of 1838 he was well prepared—indeed a cynic might be tempted to say predisposed—to appreciate the struggle for existence. But what precisely did he glean from Malthus? Historians have differed widely over the extent to which Darwin really was indebted to the celebrated essay on population. As we saw in Unit 12, he interpreted Malthus very differently from either Paley or Malthus himself, and, as Coleman argues, it was perhaps more the image of competition than a reasoned argument that Darwin assimilated. If he already had a 'theory' did he glean anything at all? Rather than involve ourselves in this difficult controversy let us pause to consider a letter which Darwin wrote to Leonard Jenyns some seven years later:

> There is one subject, on which I am very curious, and which perhaps you may throw some light on . . . namely, what are the checks and what the periods of life,—by which the increase of any given species is limited. Just calculate the increase of any bird, if you assume that only half the young are reared, and these breed: within the *natural* (i.e. free from accidents) life of the parents the number of individuals will become enormous, and I have been much surprised to think how great destruction *must* annually or occasionally be falling on every species, yet the means and period of such destruction is scarcely perceived by us.[2]

What can you find in this paragraph that is reminiscent of Malthus?

There are two things one might mention: the use of the word 'checks', and the appeal

[1] F. Darwin (ed.), *The Life and Letters of Charles Darwin*, vol. II, London, 1887, pp. 5–9; and in a more authoritative and comprehensive transcription: G. de Beer, *Bulletin of the British Museum*, Historical Series, 1960, **2**, pp. 23–200; 1967, **3**, pp. 129–76.

[2] F. Darwin (ed.), *op. cit.*, vol. II, p. 32.

to a calculation, the result of which confirms that destruction *must* be rampant even though we may rarely observe it.

Did you also notice Darwin's sense of amazement—'I have been much surprised to think'—which might just indicate that he had been equally amazed when he had first read Malthus? Already aware of a struggle in nature, he had perhaps in 1838 not fully appreciated that crucial word *must* which he was later to italicize in the letter to Jenyns. There is of course an ambiguity in the word 'must'. It could be used simply to refer to a necessary conclusion drawn from a comparison of the actual population of a given species with what the population would have been, had its increase not been checked. But this conclusion was undoubtedly reinforced by—derived its prescriptive power from—the consideration that there would always be inadequate food resources to support a geometrically increasing population. A pair of elephants—not the world's fastest breeders—could, so Darwin calculated, effectively produce 15 million progeny during the space of a mere five centuries. Considerable extermination must continue to be inevitable. If large scale extermination was constantly and necessarily taking place then even a minute differential could be of decisive significance. If such minute variations could then accumulate over the course of thousands, even millions of years, could one species gradually be transformed into another? Competition between the advanced and the normal members of a given population would mean that in the course of time the original stock might well become extinct and many if not all the transition forms also. The enormous wastage of Nature was the very motor that a process of evolution required.

This was the germ of the theory of evolution by natural selection, the first full draft of which Darwin had prepared by 1844—not for immediate publication, but certainly as a precaution lest the theory should die with him. To his wife, Emma, he wrote on 5 July 1844: 'If, as I believe, my theory in time be accepted even by one competent judge, it will be a considerable step in science. I therefore write this in case of my sudden death, as my most solemn and last request, which I am sure you will consider the same as if legally entered in my will, that you will devote £400 to its publication. . . .' Darwin still had thirty-eight years to live, but a preoccupation with his own ill-health was characteristic of almost every letter he wrote. Scholars have attributed his malaise to sources as diverse as Victorian melancholy, neuroses of a religious or paternal origin, and the bite of a South American bug. A recent and

Figure 7 Charles Darwin's study, Down House. (By kind permission of the President and Council of the Royal College of Surgeons.)

very plausible explanation is that Darwin was unwittingly but systematically poisoning himself by taking Fowler's medicine—which contained arsenic.[1] Whatever the cause, he lived at Down House in relative retirement from September 1842, protecting himself from the flurries and the palpitations of which he had complained to Henslow as far back as October 1837.

In the seclusion of his study the theory took shape, but it would be a great mistake to suppose that in the 1844 essay Darwin had solved all his problems, anticipated all objections, or even satisfied himself about every element that was later to be incorporated into the *Origin.* One such element was the effect of isolation. As he wrote to Joseph Hooker, the distinguished botanist, in that same year, 1844:

> The conclusion, which I have come at is, that those areas, in which species are most numerous, have oftenest been divided and isolated from other areas, united and again divided, a process implying antiquity and some changes in the external conditions. This will justly sound very hypothetical. I cannot give my reasons in detail; but the most general conclusion, which the geographical distribution of all organic beings, appears to me to indicate, is that isolation is the chief concomitant or cause of the appearance of *new* forms (I well know there are some staring exceptions).[2]

How would Darwin apply this insight to the Galapagos data? Given that the climate and environment of each island were essentially the same, how could the flora and fauna of the Galapagos islands be used to embarrass the theory of Lamarck (cf. Unit 12)?

The isolation of the Galapagos islands from the mainland of South America would explain why, despite the affinities between mainland and island species, there were also what Darwin called 'aboriginal' species. A peculiar variation once having occurred within an isolated region would, if cumulative, inevitably produce species peculiar to that region. This same insight would also explain the diversity of species *within* the archipelago. The fact that several islands had their own indigenous species could be explained by postulating the effective isolation of one island from another. (Darwin himself attributed this isolation to the strong currents running between the islands, to the absence of gales which might otherwise blow birds, insects and seeds from one island to another, and to the geologically recent volcanic origins of the archipelago which suggested that the islands might never have been united.) Lamarck had argued that deviations from a regular progression in the transformation of species were occasioned by responses to a changing environment. Since the environment provided by each island was the same, one would expect, on Lamarck's view, identical species on identical islands. This is not what Darwin had found.

The importance of isolation had been recognized by Darwin back in 1837, and his confidence in the letter to Hooker might suggest that he was ready to announce his convictions to the world. Note, however, the following letter, also written to Hooker in the same year, 1844:

> I forget my last letter, but it must have been a very silly one, as it seems I gave my notion of the number of species being in great degree governed by the degree to which the area had been often isolated and divided; I must have been cracked to have written it, for I have no evidence, without a person be willing to admit all my views, and then it does follow; but in my most sanguine moments, all I expect, is that I shall be able to show even to sound naturalists, that there are two sides to the question of the immutability of species.[3]

What do you conclude from this extract?

---

[1] John H. Winslow, *Darwin's Victorian Malady*, American Philosophical Society, 1971.
[2] F. Darwin (ed.), *op. cit.*, vol. II, p. 28.
[3] F. Darwin (ed.), *op. cit.*, vol. II, p. 29.

Evidently Darwin's confidence, and indeed his intentions, are still fairly restrained. He recognizes that the argument from isolation would only make sense to someone prepared to accept his entire theoretical framework. There was no factual evidence that was not theory-laden; and even at his most optimistic he felt he could not prove his case: the most that could be achieved was to show that there was a plausible alternative to the hypothesis of separate creation.

---

5 The Pursuit of Variation

The fact that his theory would have to be accepted *in toto* or not at all, undoubtedly weighed heavily upon him. The argument from isolation would only work if variations could be cumulative. And variations could only be cumulative if they could be inherited. And even if they could be inherited that was not a sufficient guarantee that they would accumulate. After all, Lyell had insisted that there were absolute limits to variation. In order to understand Darwin's position it is therefore crucial to recognize the great importance which he attached to the study of domestic breeding. In that second letter to Hooker he proceeded to review the arguments of those who have often been hailed as his precursors: Lamarck, St Hilaire, and even the German Idealist philosophers. But 'all these absurd views arise from no one having as far as I know, approached the subject on the side of variation under domestication. . . .'[1] In the pursuit of this variation Darwin perceived the originality of his own approach. The first part of his 1844 essay, and the first chapter of the *Origin* were devoted to the power of human selection in domestic breeding, Darwin insisting that knowledge of variation under domestication afforded the 'best and safest clue' to the means of modification and co-adaptation.[2] This explains why the authorities to whom Darwin was to appeal in the *Origin* included more gardeners and pigeon fanciers than philosophers. Darwin himself became a member of two pigeon-fancying clubs and proceeded to engage in breeding experiments. In a letter of May, 1855, he declared:

> The chief point which I am, and have been for years, very curious about, is to ascertain whether the *young* of our domestic breeds differ as much from each other as do their parents . . . I have got my fantails and pouters (choice birds, I hope, as I paid 20s. for each pair . . .) in a grand cage and pigeonhouse, and they are a decided amusement to me. . . .

Why should Darwin wish to answer this particular question?

One might suggest that it was to satisfy himself that, in some cases, a parental characteristic could be transmitted to the next generation without being weakened by the blending characteristic of reproduction.

It is interesting that Darwin should italicize the *young*, because there was undoubtedly another question at the back of his mind. You can perhaps infer what it was from the following passage that was later to appear in the *Origin*:

[1] F. Darwin (ed.), *op. cit.*, vol. II, p. 29.
[2] Charles Darwin, *On the Origin of Species by means of Natural Selection*, Thinkers Library Reprint of the first edition, London 1950, Introduction, p. 3.

> As we see that those variations which under domestication appear at any particular period of life tend to reappear in the offspring at the same period—for instance . . . in the eggs of poultry, and in the colour of the down of their chickens; in the horns of our sheep and cattle when nearly adult—so in a state of nature, natural selection will be enabled to act on and modify organic beings at any age, by the accumulation of profitable variations at that age, and by their inheritance at a corresponding age.

What can you infer from this extract about Darwin's interest in variation under domestication?

One might infer that breeding experiments would help to answer the question whether a given variation under domestication did in fact appear in the young—the offspring—at precisely the same stage in its development as in its parent. If so, then natural selection would be all the more potent because it would be operative and effective at every point in the life of plants, animals and their offspring.

Darwin was never to know the cause of variation, but he suspected that it must have something to do with 'the male and female reproductive elements having been affected prior to the act of conception'.[1] He also knew that a variation in one part of an organism would almost certainly be accompanied by variation in other parts. There was a curious correlation of parts: cats with blue eyes were invariably deaf; pigeons with short beaks had small feet, with long beaks large feet. But how *much* variation was possible? How *much* could be accumulated by the selective breeder? In the case of pigeons Darwin was undoubtedly astonished by the diversity of breed:

Figure 8 Artificial selection illustrated with reference to various breeds of pigeons: frillbacks (top) and fantails. Engraving by the Leighton brothers (courtesy of the Racing Pigeon Pictorial).

> Altogether at least a score of pigeons might be chosen which if shown to an ornithologist, and he were told that they were wild birds, would certainly, I think, be ranked by him as well-defined species. Moreover, I do not believe that any ornithologist would place the English carrier, the short-faced tumbler, the runt, the barb, pouter, and fantail in the same genus . . . [yet] I am fully convinced that the common opinion of naturalists is correct, namely, that all have descended from the rock-pigeon.[2]

What are the implications of this passage?

That there can be such enormous differences between domestic varieties suggests that there is no absolute dividing line between species and varieties. Since all domestic breeds have been derived from an original species, the implication is that different species in the wild may have descended from common ancestors (cf. the discussion of species in Unit 12, Section 2.3).

In Darwin naturalist and breeder were combined in creative synthesis. His fellow breeders won their prizes by selecting such slight differences that their mind's eye was closed to the effect of accumulation during successive generations—closed to the possibility of common descent. His fellow naturalists, preoccupied with systems of classification, had often attached more importance to resemblances than to differences. Variation was, to Darwin the systematist, odious too, but, as he confided to Hooker when discussing his barnacles, it was pleasing to Darwin the speculatist. His originality lay in his transcending disciplines: the naturalist of the *Beagle* brought new questions and a profound sense of perspective to the study of domestic variation, and the result was a fertile theory of natural selection.

[1] Charles Darwin, *op. cit.*, p. 7.
[2] Charles Darwin, *op. cit.*, p. 19.

6 A Controversial Analogy

Figure 9 Charles Darwin in 1860 (Down House. By kind permission of the President and Council of the Royal College of Surgeons).

The very choice of the metaphor 'natural selection' is testimony to Darwin's concern with selective breeding. If descent with modification occurred in the wild, then nature must have her own powers of selection. But was it enough to say that in the struggle for existence, favourable variations would be preserved, unfavourable ones destroyed? How close was the analogy between human selection and natural selection? Could the analogy be sustained? What were the problems associated with it? How did Darwin resolve them? These are such important questions that you must study the way in which Darwin presented the analogy in *On the Origin of Species*. The following extracts are taken from the first edition. Examine them carefully, paying particular attention to Darwin's exploitation of the analogy, and his resolution of the problems with which it is associated:

> Having alluded to the subject of reversion, I may here refer to a statement often made by naturalists—namely, that our domestic varieties, when run wild, gradually but certainly revert in character to their aboriginal stocks. Hence it has been argued that no deductions can be drawn from domestic races to species in a state of nature. I have in vain endeavoured to discover on what decisive facts the above statement has so often and so boldly been made. . . .
>
> It seems to me not improbable that if we could succeed in naturalizing, or were to cultivate during many generations the several races, for instance, of the cabbage, in very poor soil . . . that they would to a large extent, or even wholly, revert to the wild aboriginal stock. Whether or not the experiment would succeed is not of great importance for our line of argument; for by the experiment itself the conditions of life are changed. If it could be shown that our domestic varieties manifested a strong tendency to reversion—that is, to lose their acquired characters, while kept under unchanged conditions, and while kept in a considerable body, so that free intercrossing might check, by blending together, any slight deviations of structure—in such case, I grant that we could deduce nothing from domestic varieties in regard to species. But there is not a shadow of evidence in favour of this view: to assert that we could not breed our cart and racehorses, long and short-horned cattle, and poultry of various breeds, and esculent vegetables, for an almost infinite number of generations, would be opposed to all experience. . . .
>
> It has often been assumed that man has chosen for domestication animals and plants having an extraordinary inherent tendency to vary, and likewise to withstand diverse climates. I do not dispute that these capacities have added largely to the value of most of our domesticated productions; but how could a savage possibly know when he first tamed an animal whether it would vary in succeeding generations and whether it would endure other climates? Has the little variability of the ass or guinea-fowl, or the small power of endurance of warmth by the reindeer, or of cold by the common camel, prevented their domestication? I cannot doubt that if other animals and plants, equal in number to our domesticated productions, and belonging to equally diverse classes and countries, were taken from a state of nature, and could be made to breed for an equal number of generations under domestication, they would vary on an average as largely as the parent species of our existing domesticated productions have varied. . . .
>
> Hairless dogs have imperfect teeth; long-haired and coarse-haired animals are apt to have, as is asserted, long or many horns; pigeons with feathered feet have skin between their outer toes; pigeons with short beaks have small feet, and those with long beaks large feet. Hence, if man goes on selecting, and thus augmenting, any peculiarity, he will almost certainly unconsciously modify other parts of the structure, owing to the mysterious laws of the correlation of growth. . . .
>
> At the present time, breeders try by methodical selection, with a distinct object in view, to make a new strain or sub-breed superior to anything existing in the country. But for our purpose, a kind of selection which may be called unconscious and which results from everyone trying to possess and breed from the best individual animals, is more important. Thus, a man who intends keeping pointers naturally tries to get as good dogs as he can, and afterwards breeds from his own best dogs, but has no wish or expectation of permanently altering the breed. Nevertheless I cannot doubt that this process continued during centuries, would improve and modify any breed. . . .
>
> If there exist savages so barbarous as never to think of the inherited character of the

offspring of their domestic animals, yet any one animal particularly useful to them, for any special purpose, would be carefully preserved during famines and other accidents, to which savages are so liable, and such choice animals would thus generally leave more offspring than the inferior ones; so that in this case there would be a kind of unconscious selection going on. We see the value set on animals even by the barbarians of Tierra del Fuego, by their killing and devouring of old women, in times of dearth, as of less value than their dogs. . . .

But probably the most important point of all is that the animal or plant should be so highly useful to man, or so much valued by him, that the closest attention should be paid even to the slightest deviation in the qualities or structure of each individual. . . .

Under domestication it may be truly said that the whole organization becomes in some degree plastic. Let it be borne in mind how infinitely complex and close-fitting are the mutual relations of all organic beings to each other and to their physical conditions of life. Can it, then, be thought improbable, seeing that variations useful to man have undoubtedly occurred, that other variations useful in some way to each being in the great and complex battle of life should sometimes occur in the course of thousands of generations? . . .

We have seen that man by selection can certainly produce great results, and can adapt organic beings to his own uses through the accumulation of slight but useful variations given to him by the hand of Nature. But natural selection, as we shall hereafter see, is a power incessantly ready for action, and is as immeasurably superior to man's feeble efforts as the works of Nature are to those of Art. . . .

[Man] does not allow the most vigorous males to struggle for the females. He does not rigidly destroy all inferior animals, but protects during each varying season, as far as lies in his power, all his productions. . . . Under nature, the slightest difference of structure or constitution may well turn the nicely-balanced scale in the struggle for life, and so be preserved. How fleeting are the wishes and efforts of man! how short his time! and consequently how poor will his products be, compared with those accumulated by nature during whole geographical periods. . . .

Unless profitable variations do occur, natural selection can do nothing. Not that, as I believe, any extreme amount of variability is necessary; as man can certainly produce great results by adding up in any given direction mere individual differences, so could Nature, but far more easily, from having incomparably longer time at her disposal. . . .

If man can in a short time give elegant carriage and beauty to his bantams, according to his standard of beauty, I can see no good reason to doubt that female birds, by selecting during thousands of generations, the most melodious or beautiful males, according to their standard of beauty, might produce a marked effect. . . .

In man's methodical selection, a breeder selects for some definite object, and free intercrossing will wholly stop his work. But when many men, without intending to alter the breed, have a nearly common standard of perfection and all try to get and breed from the best animals, much improvement and modification surely but slowly follow from this unconscious process of selection, notwithstanding a large amount of crossing with inferior animals. Thus it will be in nature; for within a confined area, with some place in its polity not so perfectly occupied as it might be, natural selection will always tend to preserve the individuals varying in the right direction, though in different degrees, so as better to fill up the unoccupied place. . . .

Each new variety of species, during the progress of its formation, will generally press hardest on its nearest kindred, and tend to exterminate them. We see the same process of extermination among our domesticated productions, through the selection of improved forms by man. Many curious instances could be given showing how quickly new breeds of cattle, sheep and other animals, and varieties of flower, take the place of older and inferior kinds. In Yorkshire it is historically known that the ancient black cattle were displaced by the long-horns, and that these 'were swept away by the short-horns' (I quote the words of an agricultural writer) 'as if by some murderous pestilence.'. . .

A fancier is struck by a pigeon having a slightly shorter beak; another fancier is struck by a pigeon having a rather longer beak; and on the acknowledged principle that 'fanciers do not and will not admire a medium standard, but like extremes', they both go on (as has actually occurred with tumbler-pigeons) choosing and breeding from birds with longer and longer beaks, or with shorter and shorter beaks. . . . Here then, we see in man's productions the action of what may be called the principle of divergence, causing differences, at first barely appreciable, steadily to increase, and the breeds to diverge in character both from each other and from their common parent. But how, it may be asked, can any analogous principle apply in nature? I believe it can and does apply most efficiently, from the simple circumstance that the more diversified the descendants from any one species become in structure, constitution, and habits, by so much will they be better enabled to seize on many and widely diversified places in the polity of nature, and so be enabled to increase in numbers.

Using the above material you should now write a short critical review of the different uses which Darwin makes of the analogy between human and natural selection. In your answer you should mention the objections which Darwin has to anticipate and any aspect of the analogy which you may consider rather strained. You ought also to explain why Darwin repeatedly applies the word 'unconscious' to human selection.

My own review would include reference to the different roles which the analogy is called upon to play. Sometimes it is used as a kind of rhetorical device—as when Darwin refers to natural selection as 'incessantly ready for action' and 'immeasurably superior to man's feeble efforts'. At other times it serves a limited but illustrative purpose—as when the Yorkshire cattle are invoked to exemplify the point that the severest competition takes place between variants and their nearest kindred. Occasionally it fulfils an explanatory function—as when Darwin articulates his principle of divergence on the premiss that 'fanciers do not and will not admire a medium standard, but like extremes'. Finally, if the metaphor is to have any credence at all, the analogy must have a constitutive role as when Darwin takes as his 'more important' analogue the 'unconscious' selection which results from everyone trying to possess and breed from the best individual animals. Nature, in her own way, did the same. Another point one might make is that Darwin does not always use the word 'selection' in the same way. In the context of the bantam he is discussing 'sexual selection' as distinct from 'natural selection.'

There are at least three objections to the analogy which Darwin has to anticipate: the reversion of domestic varieties when transferred to the wild; the exceptional variability of domestic species; and the retrogression in nature which must follow from superior variants back crossing with inferior animals. This last objection he anticipates by insisting that such deleterious crossing must often occur under domestication, yet without preventing a trend towards improvement. The second objection relies, according to Darwin, on ignorance: how do we know that other wild species would not variegate if brought under domestication? But it was the first objection that was superficially the most powerful, since, as Darwin acknowledged, it had been used to undermine the analogy altogether. His rejoinder was ingenious: reversion might occur on transferring domestic varieties to the wild but such a transfer entailed an abrupt change in the conditions of life—a change which had no necessary analogue in the case of creatures already in the wild. If the appropriate conditions in nature were maintained there was no reason why natural varieties would have to revert to a former species.

Aspects of the analogy which you may have considered rather strained might be these: in domestic breeding, variations are selected for the benefit of man, but in nature the benefit has to be for the animal itself; secondly, despite Darwin's category of 'unconscious' selection under domestication, the fact remains that such selection still involved voluntary interference on man's part, whereas it would make little sense to ascribe voluntary action to Nature. Darwin was obviously aware of this problem, otherwise he would not have pressed the word 'unconscious', but it could be argued that, in the context of domestic breeding, his contrast should have been between methodical and undeliberate rather than between methodical and 'unconscious'.

This last criticism does not of course destroy Darwin's argument, but that it can be formulated at all is sufficient to explain why Darwin could be so readily misunderstood. As Robert Young has pointed out, the voluntaristic aspect of the analogy could be seized to justify a theistic interpretation of 'natural selection'.[1] If man did the selecting through domestication, why should God not do the selecting through

[1] R. M. Young, 'Darwin's Metaphor: Does Nature Select?', *The Monist*, 1971, **55**, 442–503.

nature? The scope for misunderstanding was so great that Darwin's colleagues would try to convince him that the phrase 'natural selection' was an infelicitous choice. In retrospect it is probably true to say that it was not so much the phrase itself, as the language with which Darwin enshrined it, that created the difficulty. It had been a problem in Aristotelian philosophy to explain how the activity of nature could be both goal-directed and yet unconscious. Darwin's metaphor, *if taken literally*, generated a similar problem. In his chapter on Natural Selection, the following account was offered:

> It may be said that natural selection is daily and hourly scrutinizing, throughout the world, every variation, even the slightest; rejecting that which is bad, preserving and adding up all that is good; silently and insensibly working, whenever and wherever opportunity offers, at the improvement of each organic being in relation to its organic and inorganic conditions of life.

Was there anything misleading here in Darwin's use of language?

His personification of natural selection would surely exacerbate problems of interpretation, rather than exclude the voluntaristic connotations.

It would be premature to pursue these problems further at this stage. You should by now be acquainted with Darwin's concept of natural selection, but you have still to see what place it occupied in the strategy of the *Origin*. In the next section you will have the opportunity to read the concluding chapter of the book for yourself, and to sample Darwin's expository technique.

7 One Long Argument

Darwin referred to his volume *On the Origin of Species* as 'one long argument'. It would have been longer still had he not been impelled towards publication by a series of events which took place in 1858. He had been working towards a much larger volume, of which the *Origin* was a mere abstract, when he was stung into activity by the realization that a young naturalist by the name of Alfred Russel Wallace had articulated a theory of natural selection that was essentially the same as his own.

Figure 10 Alfred Russel Wallace, aged 24 (1848), from a daguerrotype (from H. L. McKinney, Wallace and Natural Selection. *Courtesy of Yale University Press).*

Wallace had, rather naïvely perhaps, been converted to the idea of 'evolution' when he read Chambers' *Vestiges* in 1845 (cf. Unit 12). Once converted he had embarked upon a systematic attempt to find the mechanism. Whereas Darwin was led to his concept of evolution as a consequence of a voyage which had given him first hand experience of the geographical distribution of species, Wallace sailed to South America, and later to the Malay Archipelago, to corroborate a preconceived idea. By 1855 he had already assembled the geological and geographical evidence in favour of evolution, and he could publish a paper with the title: *On the Law Which has Regulated the Introduction of New Species*—a paper which was known to both Lyell and Darwin, and which prompted them both to attack the species problem with new zeal. Lyell opened his Species Journal in order to evaluate Wallace's position; Darwin, in May 1856, began what he called his Species sketch to take his own arguments further. A year earlier Darwin had been resigned to collecting all the facts that he could muster 'to see how far they favour or are opposed to the notion that wild species are mutable or immutable: I mean with my utmost power to give all arguments and facts on both sides'.[1]

Now, albeit reluctantly, his writing had to take first place: 'I rather hate the idea of writing for priority', he informed Lyell, 'yet I certainly should be vexed if any one were to publish my doctrines before me.'[2]

Since both Lyell and Darwin were affected by Wallace's paper of 1855 you might like to examine a summary of the evidence which Wallace had in fact adduced in favour of his law:

Wallace's Summary of the Geological and Geographical Evidence in His 1855 Paper[3]

GEOGRAPHY

1. Large groups, such as classes and orders, are generally spread over the whole earth, while smaller ones, such as families and genera, are frequently confined to one portion, often to a very limited district.

2. In widely distributed families the genera are often limited in range; in widely distributed genera, well-marked groups of species are peculiar to each geographical district.

3. When a group is confined to one district, and is rich in species, it is almost invariably the case that the most closely allied species are found in the same locality or in closely adjoining localities, and that therefore the natural sequence of the species by affinity is also geographical.

4. In countries of a similar climate, but separated by a wide sea or lofty mountains, the families, genera and species of the one are often represented by closely allied families, genera and species peculiar to the other.

GEOLOGY

5. The distribution of the organic world in time is very similar to its present distribution in space.

6. Most of the larger and some small groups extend through several geological periods.

7. In each period, however, there are peculiar groups, found nowhere else, and extending through one or several formations.

8. Species of one genus, or genera of one family occurring in the same geological time are more closely allied than those separated in time.

9. As generally in geography no species or genus occurs in two very distant localities without also being found in intermediate places, so in geology the life of a species or genus has not been interrupted. In other words, no group or species has come into existence twice.

10. The following law may be deduced from these facts—Every species has come into existence coincident both in space and time with a pre-existing closely allied species.

Wallace had come to the same conclusion as Darwin, but there was nothing in his 1855 paper to suggest that he had yet found a mechanism, let alone that of natural selection, which would explain why, in a given locality, 'new species have been created . . . on the plan of the pre-existing ones'. His law explained the facts of geographical distribution, but the law itself was not explained. Three years later, in February 1858, it was. As with Darwin twenty years before, it was the reading of Malthus that propelled Wallace towards the concept of natural selection. He had read, and been impressed by, the *Essay* once before, but on a second reading, the chapters which dealt with the causes restricting the population of savage and barbarous tribes, had a cumulative effect on his mind. Of the American Indians Malthus had written: 'like beasts of prey, they must either drive away or fly from every rival, and be engaged in perpetual contests with each other.' It was passages like this, Wallace later recalled, that had stimulated his imagination.[4] The idea of perpetual contest, projected on to all the creatures of the globe, gave him the insight he required:

[1] F. Darwin (ed.), *op. cit.*, vol. II, p. 49.
[2] F. Darwin (ed.), *op. cit.*, vol. II, p. 68.
[3] H. L. McKinney, *Wallace and Natural Selection*, Yale University Press, 1972, p. 159.
[4] McKinney, *op. cit.*, pp. 164–5.

there was a general principle in nature which would 'cause many *varieties* to survive the parent species, and to give rise to successive variations departing further and further from the original type'. On the island of Gilolo, Wallace composed a paper which expounded his new insight. Its title was *On the Tendency of Varieties to depart indefinitely from the Original Type*, and a copy was despatched across the world to Darwin.

'I never saw a more striking coincidence', Darwin wrote to Lyell. 'If Wallace had my MS. sketch written out in 1842, he could not have made a better short abstract! Even his terms now stand as heads to my chapters.' With mock magnanimity Darwin volunteered to Lyell that he would wish to have Wallace's paper published: 'so all my originality . . . will be smashed.'[1] He had the consolation that his own book would not be depreciated since 'all the labour consists in the application of the theory', but he suffered all the anguish of one forestalled with a vengeance. The solution is well known: with the assistance of Lyell and Hooker, Darwin's claim to priority was presented to a meeting of the Linnean Society on 1 July 1858, when excerpts from Darwin's work as well as Wallace's paper were read. But it is no longer clear that the solution was quite so gentlemanlike as has usually been supposed. Wallace's paper took second place at the meeting, he never saw page proofs before publication nor learned what had happened until all was over. Scholars sympathetic to Wallace's achievement, especially if they suspect that Darwin may have learned something about divergence of character from Wallace himself, may agree with Lewis McKinney that: 'one must ask why it was so imperative that the joint papers be presented with such celerity if Darwin's priority were not being protected . . . Darwin's half-hearted pleas that he would take a back seat to Wallace cannot be taken seriously.'[2]

Be this as it may, Darwin's hunch that his fortunes would depend on his book was surely a sound one. There his theory would be applied, and used to anticipate objections—which was more than Wallace could execute in one short paper. The abstract of Darwin's original book was to be published as a series of papers in the *Journal of the Linnean Society*, but even his abstract grew too big and it was eventually published in one volume on 12 November 1859, all 1,250 copies selling that very day.

The concluding chapter of the *Origin* is reproduced in *Anthology I*, and now is the time to read it. Since this lies at the heart of the unit you should study it with great care. You will find that Darwin enumerates a long series of facts which he considers to be explained by his own theory but not explained by a theory of separate creation. You should tabulate these facts for your own consideration as you proceed. When you have finished doing that, you should answer these further questions:

---

1 In an earlier chapter of the book Darwin had insisted that in contemplating the history of organic life our *reason* must conquer our *imagination*. Why would he have done so?

2 Early in this concluding chapter he mentions a case which poses a special difficulty to his theory of natural selection – the case of the caste system among sterile female ants. Can you work out what the difficulty was and how Darwin might have dealt with it?

3 What difficulty does the general distribution of animals present?

4 When Darwin refers to the absence of connecting links or intermediate varieties, he could be referring to one of two different objections. What were they, and which was the most serious? How does he cope with it?

5 Why would sexual selection be especially important in Darwin's theory?

6 Why is there a constant tendency in natural selection to preserve the most divergent offspring of any one species?

7 Is anything in Darwin's discussion reminiscent of Larmarck?

8 To what did Darwin attribute the persistence of belief in immutable species?

9 Is it a necessary part of Darwin's theory that all life has evolved from a single prototype?

10 What did Darwin consider his basic contribution to be?

11 Did he conceive his theory to be essentially anti-religious?

12 In what sense is Darwin's conclusion optimistic? How does it compare with that of Malthus?

---

[1] F. Darwin (ed.), *op. cit.*, vol. II, pp. 116–17.

[2] McKinney, *op. cit.*, p. 143.

1 The difficulty of explaining how the more complex organs and instincts have come to be perfected was one which, in the chapter you have just read, Darwin described as 'appearing to our imagination insuperably great'. However, given that there could be gradations in the perfection of any organ or instinct, 'each good of its kind', the gradual perfection of an eye, for example, would not be offensive to reason.

2 The caste system among sterile ants posed a problem because the action of natural selection required the inheritance of variation, but sterile ants could not transmit the characteristics of their caste to the next generation. To solve the problem Darwin could apply his concept of natural selection to the ant community, for the survival of which it was evidently an advantage to contain neuters of different kinds. By the long-continued selection of fertile parents which produced the most neuters with a profitable modification, so all the neuters would eventually acquire characteristics that were favourable but not necessarily identical. The preservation of two or three different modifications could lead to a caste system of two or three.

3 Though the geographical distribution of animals provided Darwin with some of his most telling evidence, he admitted that he was 'often wholly unable even to conjecture' how individuals of the same species, or species of the same genus, could have spread to such distant and isolated parts of the world. He does, nevertheless, proceed to argue that species might have remained stable long enough for wide migration to occur.

4 One objection could be the absence of connecting links among existing forms; the other, and more forcible, their absence in the fossil record. The latter was the more serious since, on the theory itself, one had no right to expect direct links between living species—only between a given species and an extinct progenitor. Darwin copes with the objection by resorting, as Lyell had done for a different motive, to the imperfection of the geological record. Precisely because intermediate forms had been intermediate, and therefore relatively transient and localized, it would be remarkable if they had been fossilized.

5 Sexual selection was so important because it ensured that 'the most vigorous individuals . . . will generally leave most progeny'. The accumulation of sexually advantageous variation would therefore be all the more probable.

6 According to Darwin, divergence is encouraged by the possibilities which it creates for descendant species to fill more and more diverse niches in the economy of nature.

7 Darwin's references to the effects of disuse, and changed habits, are reminiscent of Lamarck. Natural selection is not presented as the exclusive factor in changing an organ.

8 Persistent belief in immutable species was attributed to the prevalence of the assumption that the geological record was sufficiently perfect to demonstrate mutation if there had been any; and to the reluctance to admit great changes when intermediate steps could not be perceived.

9 Darwin himself clearly favours a single primordial form, but the fact that he can say: 'I believe that animals have descended from at most only four or five progenitors . . .' would suggest that he was prepared to tolerate a few more.

10 New fields of enquiry would open up in all directions, but Darwin undoubtedly felt that his chief contribution was to slay the ghost of *essential* species. No longer need taxonomists be haunted by the old question whether their specimens were truly species: in the last analysis the concept of species was just a convenient device, though some were clearly more well-defined than others.

11 That he had been so reluctant to commit his theory to print suggests that he was sensitive to the anti-religious constructions that might be placed upon it. At the same time, he did not propound the theory as if it had atheistic implications. On the contrary, the postulation of natural laws and the continuity of secondary causes was consistent with belief in a Creator who had impressed these laws on matter (cf. Professor Hooykaas' discussion of deism in Unit 11).

12 There is an optimistic strand in Darwin's conclusion which follows from the postulate that 'natural selection works solely by and for the good of each being'.

Consequently 'all corporeal and mental endowments will tend to progress towards perfection'. Ironically this was the very belief that Malthus had set out to destroy in the first edition of his essay. His image of nature had become incorporated into a new synthesis that promised the very progress that he had eschewed.

A more thorough investigation of Darwin's significance would have to develop each of these points in turn. There is little doubt, for example, that it was the optimistic strand in his conclusion that made his theory palatable to social and political theorists who were looking for a message of hope. Here was scientific sanction for the inevitable progress which Herbert Spencer had already projected into the future. We shall return to this theme later, but for the moment we ought to remain with some of the obstacles to belief in natural selection.

Darwin was arguably at his best when anticipating objections to his theory. He was scrupulously honest, and he had, after all, had twenty years to come to terms with them. How could a land carnivore be converted to one with aquatic habits when transitional forms might not be able to subsist? How had the bat developed its wings, when they would have had to be more than rudimentary to have conferred any advantage on a distant precursor? Were there not other cases of organs which could be of no use until fully formed? Darwin thought not: each small modification in the development of an organ could be useful. With the encyclopaedic knowledge of a brilliant naturalist, he could always produce a salient example to forestall his critics. 'Look at the family of squirrels; here we have the finest gradation from animals with their tails only slightly flattened, and from others . . . with the posterior part of their bodies rather wide and with the skin on their flanks rather full, to the so-called flying squirrels . . . we cannot doubt that each structure is of use to each kind of squirrel in its own country, by enabling it to escape birds or beasts of prey, or to collect food more quickly.'

With most objections Darwin could cope. But how definitive was his long argument? What problems remained outstanding? Were all scientists of repute immediately convinced? The fact is that partly as a result of scientific criticism, and mainly as a result of his own further enquiries, Darwin's argument was to become much attenuated in successive editions. The primacy of natural selection, which was later incorporated within neo-Darwinian orthodoxy, was effectively sacrificed by Darwin himself. This is of course an aspect of the evolutionary debate which is quietly ignored or lamented by those who like to record the linear triumph of truth. Nevertheless in the *Descent of Man* (1871), Darwin freely admitted 'that in the earlier editions of my *Origin of Species* I probably attributed too much to the action of natural selection or the survival of the fittest . . . I had not formerly sufficiently considered the existence of many structures which appear to be, as far as we can judge, neither beneficial or injurious; and this I believe to be one of the greatest oversights as yet detected in my work'.

During the twelve years which separated the *Descent* from the *Origin*, Darwin was accused of many oversights—sometimes justly, sometimes unjustly—and he was forced into 'documented qualification and nagging doubt'.[1]

In the *Origin*, natural selection worked on minute 'individual differences', not on sudden mutations—or saltations as they were called (from the latin 'saltus', meaning a 'leap'). It was therefore possible to complain, as the Harvard geologist and zoologist Agassiz did, that Darwin's individual differences were merely individual peculiarities, transient and not inheritable. Among his allies, T. H. Huxley felt that Darwin had sabotaged his case by his very rejection of saltations: the action of natural selection on more substantial variations would surely have the great advantage of accelerating the process of speciation? According to another correspondent, William Henry Harvey, the adoption of saltations would have further merit: 'by admitting the possibility of *sudden* divarications we get rid of those perfectly innumerable forms of life which your hypothesis requires us to believe in, but of whose existence

[1] P. J. Vorzimmer, *Charles Darwin: The Years of Controversy*, University of London Press, 1972, p. 20.

there is so little evidence either in existing nature or among fossils.'[1] Darwin conceded little on this issue, especially after reading Fleeming Jenkin's celebrated review of the *Origin*'s fourth edition.[2] In his article in the *North British Review* (1867), Jenkin argued that any single saltation occurring in an animal population was bound to be swamped by breeding with the original stock. Darwin, who shared the prevailing opinion that the most prominent feature of reproduction was a *blending* of parental characteristics, was bound to agree. Darwin knew there were exceptions to the blending principle. Parental characteristics had been known to recur at several generations' remove. But these were the exceptions. Natural selection could not work on isolated saltations, because any beneficial variation would be diluted rather than accumulated. But could the same argument not be applied to his cherished small differences? This was another problem which the theory had to face: if the raw materials for natural selection were *individual* differences, would they not also be swamped in a large population?

Darwin had been worried by this before, and had made several moves towards solving the problem. If a small number of variants had to retain their integrity, then perhaps some form of isolation might 'protect' them from a large part of the general pool? But if this argument were to work, the isolated populations would have to be small, and this would create more problems than it solved. The more restricted the population, the less chance was there of favourable variation occurring, and the less chance of the severe competition required for selection. Had Darwin had the genetics of Mendel (1822–84) at his disposal, his dilemma would have been less acute, since Mendel's ill-fated association of particular characteristics with genetic units that could be transmitted through successive generations (1865), was much later invoked to countenance the persistence and not the swamping of variation. Without that genetic model, Darwin had to load the dice against swamping in some other way.

Can you suggest any way in which this might be done?

Since the swamping argument presupposed a large disproportion between the number of variants and the total population, another move would be to increase the proportion of effective variations. There were at least two ways in which Darwin could achieve this. In the *Origin* itself he had intensified the destruction in nature to such a degree that under changed conditions almost all organisms would be destroyed except the few minimally adapted. The alternative was to argue that there were good reasons why a particular variation might occur among larger numbers of a given population at any one time.

The former explanation required no more than the intensive power of natural selection but the latter opened the door to all kinds of possibilities which seemed to detract from that very power. The hypothesis of simultaneous variation had the attraction that it allowed Darwin to turn blending inheritance to his advantage. Given a sufficient number of better adapted variants within a community, then blending with the pool would help to distribute the benefit. But how could similar variations be simultaneously induced in the first place? Ignorant of the source of variation Darwin could only conjecture, but his conjectures were apt to appear destructive of his own originality.

Perhaps if variations were induced by external conditions that were common to a given population, one would *expect* similar variations to occur simultaneously? As he pursued the causes of variation he was constrained to stress the direct effect of physical conditions, and to attach such increasing weight to the effects of habit, use and disuse, as would have flattered Lamarck. Studiously separating the causes of variation from the causes of evolution, Darwin did his best to protect his mechanism of natural selection; but a letter to Hooker of 1862 reveals his innermost feelings:

[1] Vorzimmer, *op. cit.*, p. 62.
[2] Vorzimmer, *op. cit.*, pp. 121–4.

I hardly know why I am a little sorry, but my present work is leading me to believe rather more in the direct action of physical condition. I presume I regret it, because it lessens the glory of natural selection, and is so confoundedly doubtful.[1]

The 'fact' of evolution was hardly in question, but no sooner had it shone, than the glory of natural selection went into partial eclipse. Its explanatory power was accepted by the inner circle of evolutionists, including Wallace and Huxley, but its sufficiency was still a matter for debate. And not only for 'scientific' reasons. As we shall see in the next section, Darwin had presented his theory of natural selection in such a manner that it could easily become grist to the philosopher's mill.

8 The Status of the Theory

By virtue of the fact that his theory was a theory about the past, Darwin was bound to speak of probabilities rather than certainties. But in so doing he was exposing himself to criticism which to us may appear trivial, but which to many of his contemporaries, and distinguished scientists among them, had to be taken seriously. Before we proceed to a more systematic account of the reception of Darwin's work, it will be instructive to examine some of these criticisms, and some of the logical difficulties associated with his argument.

From the extracts reproduced below can you deduce one source of objections that were likely to be levelled at the *Origin*?

> To be brief, I assume that species arise like our domestic varieties with *much* extinction; and then test this hypothesis by comparison with as many general and pretty well-established propositions as I can find made out—in geographical distribution, geological history, affinities, etc. etc. And it seems to me that *supposing* that such hypothesis were to explain such general propositions, we ought, in accordance with the common way of following all sciences, to admit it till some better hypothesis be found out (DARWIN).[2]
>
> We can clearly see why these animals which cannot cross wide spaces of ocean, as frogs and terrestrial mammals, should not inhabit oceanic islands; and why, on the other hand, new and peculiar species of bats, which can traverse the ocean, should so often be found on islands far distant from any continent. Such facts as the presence of peculiar species of bats, and the absence of all other mammals on oceanic islands, are utterly inexplicable on the theory of independent acts of creation (DARWIN).[3]
>
> I cannot as yet (probably never shall) receive the theory of natural selection as a satisfying explanation of the origin of species—but I am willing to admit that it explains several facts which are not otherwise easily to be accounted for. Until, however, something more is known of the inciting causes of the variations and correlations of organs I can only regard natural selection as one agent out of several (WILLIAM HARVEY).[4]

Can you state what logical problem might well have opened the floodgates to criticism?

The problem was that natural selection could be dismissed as one hypothesis among several that might explain the facts. Darwin admitted in the letter to Gray that he was testing a hypothesis—testing an assumption—that might prove the best available. Conservative minds, however, were unlikely to be impressed by Darwin's rhetoric

[1] Vorzimmer, *op. cit.*, p. 82.
[2] F. Darwin (ed.), *op. cit.*, vol. II, Darwin to Asa Gray, p. 78.
[3] Charles Darwin, *op. cit.*, p. 405.
[4] Vorzimmer, *op. cit.*, William Harvey to Darwin, p. 60.

against the hypothesis of independent acts of creation. One can sympathize with his conviction that no Intelligence could have had a sufficient reason for depositing only bats on an oceanic island, but there was surely no sufficient reason why an Intelligence should not! Moreover, as the letter from Harvey shows, natural selection and independent acts of creation were not the only possible hypotheses. Other agents might have been at work.

Taking full advantage of this logical problem, Darwin's opponents could attack him on several fronts. The following accusations were fairly typical, and you should note that some came from physical scientists of distinction:

A

'Hypotheses non fingo',
Sir Isaac Newton said
And that was true, by Jingo!
As proof demonstrat*ed*
But Darwin's speculation
Is of another sort
'Tis one which demonstration
In nowise doth support.

(*Punch* 69, 1871, 8 April, 145).[1]

B

There are many cases, indeed, in the history of science, where speculations, like those of Kepler, have led to great discoveries. . . . It is otherwise, however, with speculations which trench upon sacred ground, and which run counter to the universal convictions of mankind, poisoning the fountains of science, and disturbing the serenity of the Christian world. Such is doubtless the tendency of Mr Darwin's work (SIR BENJAMIN BRODIE).[2]

C

We are not denying our author's right to reason upon his hypothesis . . . [But] we had not dreamt that because the objections to a theory could not be proved to be absolutely insuperable we were called upon to accept it as true. We had fancied that the laws of reasoning in such matters, to which Newton and Laplace, Fresnel and Faraday, in their mathematical or experimental investigations, have bowed in reverential obedience, were still in force . . . [Darwin] constantly speaks of his theory as explaining certain phenomena, which he represents as inexplicable on any other theory. We altogether demur at this. . . . A phenomenon is properly said to be explained, more or less perfectly, when it can be clearly referred to some recognized cause . . . But Mr Darwin's theory can explain nothing in this sense, because it cannot possibly assign any necessary relation between phenomena and the causes to which it refers them (HOPKINS).[3]

D

The essence of science consists in inferring antecedent conditions and anticipating future evolutions from phenomena which have actually come under observation. In biology the difficulties of successfully acting up to this ideal are prodigious . . . I have always felt that the hypothesis of 'the origin of species through natural selection' does not contain the true theory of evolution, if evolution there has been, in biology. . . . I feel convinced that the argument of design has been greatly too much lost sight of in recent zoological speculations (WILLIAM THOMSON).[4]

E

What is there unreasonable in the supposition that they [all the animals in the world] have all been formed on the same general plan? Mr Darwin's only objection is, that 'this is not scientific explanation'. But this is simply to beg the question. If Mr Darwin starts with the preliminary assumption that every fact in Nature is capable of scientific explanation—in other words, that no causes have ever operated except natural causes, he will, of course, reject any other causes. But this assumption is the very point to be proved. To argue from it is to assume the whole doctrine of Evolution (*Times* review of *Descent of Man*).[5]

[1] A. Ellegard, *Darwin and the General Reader*, Göteberg, 1958, p. 190.

[2] A. Ellegard, 'Darwin's Theory and Nineteenth-Century Philosophies of Science', in P. Wiener and A. Noland (eds.), *Roots of Scientific Thought*, Basic Books, New York, 1957, pp. 560–1.

[3] Cited in Ellegard, *ibid.*, 562.

[4] Ellegard, *ibid.*, 562.

[5] Ellegard, *ibid.*, 564.

Each of these accusations was designed to make a philosophical point. Can you extract the point in each case?

A The point here is that the scientist should refrain from making any undemonstrable hypotheses.

B Reducing Darwin's hypothesis to a mere speculation, Brodie argues that no hypothesis can be admitted which contradicts already well-established beliefs.

C Hopkins has no objection to the use of a hypothesis, but insists that to remove difficulties from a theory (as Darwin had patiently done) is not the same as establishing its truth. There is the additional implication that Darwin's theory does not provide a properly causal explanation.

D Thomson recognizes that explanations in biology are almost bound to fall short of a scientific ideal, but hints that Darwin's theory is defective for the additional reason that it ignores the argument from design and, by implication, the primacy of final causes.

E The point here is that judged from a philosophy which admits supernatural power, there is an element of circularity in Darwin's position. If continuity of natural causes is presupposed in the premisses, then it will reappear in the conclusion.

One does not have to read between the lines to detect a common fear running through most of these criticisms. Can you spell out what it was?

It is fairly clear that Darwin's theory is seen as a threat to traditional Christian beliefs.

Some of the above accusations were more serious than others. How do you think Darwin and his 'bulldog' Huxley would have answered each one?

A The above letter of Darwin to Asa Gray indicates that he knew full well that the testing of hypotheses was part and parcel of the scientific enterprise. (In fact the call to avoid hypotheses altogether would hardly have come from anyone with experience of scientific research—and the appeal to Newton relied on a misunderstanding of what Newton had meant!)

B Darwin would presumably prefer to question cherished beliefs than abandon fruitful hypotheses.

C With the first part of Hopkins' objection, Darwin would probably have agreed, but he had not made the kind of truth-claim that is imputed to him. As an empiricist, he would have felt that Hopkins' demand for necessary relations between phenomena was too metaphysical.

D If his theory had dispensed with the primacy of final causes, this was not an objection Darwin was likely to take seriously. He would obviously have considered the possibility of reintroducing them into his scheme (as we shall see later), but the view, for example, that each variation had been providentially arranged seemed to Darwin to make natural selection superfluous.[1]

E This accusation of circularity was one that Darwin could hardly deny, but he would undoubtedly have insisted that a coherent naturalistic explanation had the great merit, as the positivist philosopher G. H. Lewes proclaimed, of bringing the question within the region of research. References to a plan of creation, or to divine *fiat* did not.

---

[1] F. Darwin (ed.), *More Letters of Charles Darwin*, London, 1903, vol. I, p. 191.

There is little doubt that Darwin had adequate rejoinders to popular objections of this kind. But the circularity urged by the *Times* review of the *Descent of Man* was not the only circularity claim with which Darwin had to contend. Increasingly unhappy about the anthropomorphic connotations of the expression 'natural selection' (cf. Section 6), he came to prefer the 'survival of the fittest' as an epigram. But the circularity was transparent, so his critics claimed: that the fittest had survived and would continue to survive was tautologically true, since what we mean by fitness is the capacity to survive! In one form or another this objection has been pressed until the present day. Thus the distinguished biologist, C. H. Waddington, has written: 'The general principle of natural selection, in fact, merely amounts to the statement that the individuals which leave most offspring are those which leave most offspring. It is a tautology.'[1] Now what we have to decide is whether there can be an answer to this objection which, had it not been so consistently championed, would appear to be little more than a verbal quibble.

As Michael Ruse[2] has observed, Darwin formulates at least two versions of the natural selection principle:

1 Organisms with useful variations have a better chance of surviving and reproducing than organisms with injurious variations.
2 Useful or favourable variations are preserved and injurious variations are rejected, or, in other words, many if not most of those with favourable variations will actually survive and reproduce, and comparatively fewer of those with injurious variations will succeed.

Study these two formulations and see if you can work out for yourself how the circularity objection might be met.

The objection only has force if the 'fittest' organisms are identified with those which are bound to survive. If, however, the 'fittest' are merely those with favourable or useful variations, the objection can be met since Darwin gives no guarantee that *all* organisms with a favourable variation will perforce survive, or *all* with an injurious one perish. To define a favourable variation as one which would give an organism a better *chance* of survival does not make Darwin's argument circular.

We have so far glanced at two different allegations of circularity, but there was a third that affected the logical status of Darwin's theory. In the very nature of the case direct evidence of transmutation was out of the question. But if there were countless transition forms either still alive or preserved in the fossil record, then the theory could be in some measure corroborated. The problem was their conspicuous absence—so conspicuous that Darwin feared that few would grant him a geological record as incomplete as he was forced to plead. With great ingenuity he used the theory itself to explain why intermediate forms had been exterminated in intermediate zones, and why, by very virtue of their being intermediates, they had been too relatively few and transient to have left their mark. But what were his contemporaries to make of a theory which explained why there was apparently so little evidence in its favour? Only as more and more missing links were unearthed would this difficulty begin to fade away.

Philosophers of science will continue to argue whether Darwin's argument was deductive or inductive, whether he had an argument at all, and whether he laid down a programme for an explanation of the evolutionary process rather than the explanation itself. But there is one point that must be stressed above all. Despite Darwin's own descent from the *Origin's* first edition, and despite the difficulties we have just explained, he had succeeded in making the idea of life's evolution scientifically

[1] Cited by M. Ruse, 'Natural Selection in the *Origin of Species*', *Studies in History and Philosophy of Science*, vol. I, 1971, p. 312.
[2] Ruse, *ibid.*, 335.

respectable. In his *Antiquity of Man* (1863) and subsequently in the tenth edition of his *Principles of Geology*, Lyell was prepared to abjure his earlier arguments against the progressive development of life; and one of Darwin's staunchest critics, the anatomist and zoologist, St George Mivart, was prepared to admit that the *body* of man had arisen from a lower species, and that natural selection might have had a hand in the process. What Mivart, as a Roman Catholic, took exception to was the sufficiency of natural selection as an explanation. Though his motives were in part religious—God had to have a part in the process—his critique of natural selection provides a telling illustration of the point that the status of Darwin's description of the past, of his account of the pattern of evolution, was far more secure than that of his mechanism. Not that the two could be completely dissociated, but, as Peter Vorzimmer has observed, Mivart's stack of arguments against natural selection 'made a formidable case'.[1] If the modifications on which natural selection was supposed to work were as minute as Darwin maintained, surely they would be virtually useless, and in any case of negligible value in the struggle for existence? Was there not evidence of potentially useful saltations? Then there were the gaping gaps in the fossil record. The 'individual differences' on which Darwin relied so heavily were generally unin-heritable, and there were surely limits to the accumulative power of selection? Nor, Mivart argued, could the origin of man's soul and his mental powers be reduced to the agency of natural selection. When, in the *Descent of Man*, Darwin himself retracted his earlier commitment to the primacy of natural selection, Mivart could exploit the situation to persuade his readers that the Darwinian theory had been virtually abandoned by its creator. Backwoodsmen among later religious apologists have been able to gloat over such a gift, but however precarious Darwin's logic, his *intuitions* about nature were, as Walter Cannon has affirmed, quite sufficient to transform the sciences of biology and geology.[2]

That they were sufficient to transform other disciplines, as well as man's concept of his place in nature, compels us to look at the reception of Darwin's theory through a wide-angle lens.

9 'God Knows what the Public will think'

Thus Darwin wrote to Wallace a few days before the *Origin* was launched. Not that he cared all that much once he could reassure himself that both Huxley and Lyell were to be on his side. It transpired that his 'old master', Lyell, was never to become a full convert, but once Huxley had published his favourable review in the *Times* (26 December 1859), Darwin began to feel more secure. What he wanted most of all was that his argument should be judged on its scientific merits, and he was always relieved when reviews abstained from religious diatribe. 'You have done an essential kindness in checking the odium theologicum in the E[dinburgh] R[eview]', he wrote to his new convert, the physiologist, W. B. Carpenter. 'It much pains all one's female relations and injures the cause.'[3]

But what did the public think, those who were not able to judge the scientific merits of the case? Published in late November, the *Origin* made a good Christmas present. A young master at Rugby, in fact one of the first school science masters, Canon J. M. Wilson, gave a copy to his father, a country cleric, whose response was hardly encouraging: 'I cannot conceive how a book can be written on the subject. We know all there is to know about it. God created plants and animals and man out

[1] Vorzimmer, *op. cit.*, p. 230.

[2] W. F. Cannon, 'Darwin's Vision in *On the Origin of Species*', in G. Levine and W. Madden (eds.), *The Art of Victorian Prose*, Oxford University Press, 1968, p. 171.

[3] F. Darwin (ed.), *op. cit.*, vol. II, p. 239.

of the ground'.[1] Not all Victorian fathers were quite so adamant, but for many of them the imagery of Milton added to the authority of *Genesis,* conspired to make Darwin's theory presumptuous in the extreme. As Darwin himself was soon to perceive, he had driven a wedge between father and son. His converts belonged, for the most part, to a younger generation of naturalists. In a celebrated book entitled *Father and Son,* Edmund Gosse beautifully described the reactions of his own father, the marine biologist P. H. Gosse. A member of the Plymouth Brethren, Gosse senior took his ascetic Biblical Christianity so seriously that, as a child, Edmund had been denied the study of Shakespeare—so corrupting was the bard. What was to be made of theories of organic evolution? After much heart-searching Gosse had hit on an ingenious solution. Suppose God were to create a tree out of nothing. If it were to be a real tree, it would have to exhibit the appearance of age, with roots, rings and bark. With a logic more gratuitous than ludicrous the elder Gosse applied the analogy to the earth itself. Suppose God had created the earth and its inhabitants in accord with a Biblical perspective. If it were to be a real world, it would have to exhibit the appearance of age—exhibit the very fossils whose existence Darwin was invoking to support his argument. With great enthusiasm Gosse went into print and eagerly awaited the acclaim of his critics. But when *Omphalos* (1857) met with a cool reception, he was bitterly distressed. Indeed, Edmund could say of his father that it was as if he had become angry with God himself for having so wretchedly failed him.

Figure 11 Father and son, 1857. Frontispiece of Edmund Gosse, Father and Son *(courtesy of William Heinemann & Co. Ltd. and Miss J. Gosse).*

Among the educated public, anger was more frequently directed towards Darwin, and for obvious reasons. The public had its own image of the scientist, but here was a scientist purveying an image of the public. The reduction of man to an animal was not to be taken sitting down. However, 'Those who oppose my theory,' Darwin exclaimed, 'by snarling and baring their teeth, only confirm thereby, their canine origins.' Apes at the London Zoo are reported to have asked each other: 'Am I my keeper's brother?'; and it seems that one had to side either with them or the angels. But the familiar anecdotes do little to enhance our understanding. You will find it far more instructive if you turn instead to Huxley's account of the reception of the *Origin* which is reproduced in *Anthology I.* Using the information to be found there, and amplified by what you have already learned, you should try the following exercise:

Write two short reviews of Darwin's book, each of 500 words—one that might have been written by an enthusiastic scientist in 1860; the other by an equally unenthusiastic cleric.

Since this was a chance for you to 'go to town' on the material we shall not spoil your enjoyment by citing specimen answers. Instead we shall simply list some of the more important points that you might have included in each review:

Your favourable review could have been based on Huxley's own reaction, since he fits the description perfectly. Such a review would have to summarize Darwin's argument, stress his originality, compare him perhaps with Newton, emphasizing his authority, and indeed his 'widespread reputation among the general public'. You could have mentioned the difficulty involved in mastering the argument, in seeing things Darwin's way, at the same time drawing attention to its explanatory power. Darwin's book could also be hailed as presenting the first theory of evolution to have scientific plausibility: it could be contrasted with earlier speculations which were either deficient in their evidence or in the mechanism proposed—or, as with Chambers' thesis, the product of an unscientific habit of mind. Reference could be made to Darwin's elimination of extraordinary agencies, and parallels drawn with Lyell's achievement in geology. One could have expressed reservations about the sufficiency of Darwin's mechanism, but at the same time have illustrated its merits as a working hypothesis. As a final gesture one might have forestalled theological objections by arguing, as Huxley does, that Darwin's theory 'has no more to do with Theism than the first book of Euclid has'.

[1] Cited by J. Y. Simpson, *Landmarks in the Struggle Between Science and Religion*, London, 1925, p. 177.

Your hostile review would have to summarize Darwin's argument, drawing attention perhaps to the great faith which he has in natural selection. You would almost certainly have appealed to Biblical authority in some form or another, arguing perhaps that the *Genesis* account of creation could not be abandoned since inextricably bound up with it were the doctrines of the Fall and, in the New Testament, Christ's atonement. According to Darwin man has risen not fallen; and how could Christ be a second Adam to the rescue, if there had never been a first! If Darwin meant to imply that man himself had evolved from a more primitive mammal then this was surely incompatible with the belief that man was a special creature, fashioned in the image of God? One might well have objected to Darwin's presumption in excluding almost all reference to divine activity, associating his name with Lamarck to whose doctrines he had assuredly added nothing of value. That he had patently deserted the inductive method of reputable science would be another allegation, and one could certainly have exploited his ignorance of the causes of variation. Darwin's use of the word 'chance' was reprehensible, indicating a subversive attack on the 'fact' of God's Providential designs. If God had made man by the process of evolution as described by Darwin—a process which only worked because of competition, wastage and death—this would be inconsistent with the fullness of His Glory. And as a final gesture one might have forestalled scientific objection by appealing to the authority of more conservative scientists, such as Whewell, Owen or Agassiz.

Figure 12 'Man is but a worm' by Linley Sambourne, from Punch's Almanack for 1882 (Mansell Collection).

By setting up the exercise in this way, we have of course generated an image of warfare between science and religious orthodoxy. Nor is there any more appropriate place to do so, since it was in the context of the Darwinian debates that the warfare historiography was consolidated (cf. Unit 1). In fact the chemist and would-be historian, J. W. Draper, was one of the speakers at the celebrated meeting of the British Association in 1860 when an audience of between 700 and 1,000 witnessed that confrontation of Bishop Wilberforce and Huxley which has come to epitomize the polarization of opinion. The Bishop is reported as having spoken 'for full half-an-hour with inimitable spirit, emptiness and unfairness . . . "crammed" up to the throat'.[1] To his jibe at Huxley's choice of ancestry, there came the retort that it were better to have an ape for an ancestor than a man who would abuse his intellect by introducing aimless rhetoric into a sphere of which he was ignorant. Darwin himself was glad he was not in Oxford, since, like the lady who fainted and had to be carried

Figure 13 Bishop Samuel Wilberforce (Mansell Collection).

[1] F. Darwin (ed.), *op. cit.*, vol. II, p. 321.

out, 'I should have been overwhelmed; with my health in its present state'. On the whole, however, he was well pleased with the outcome: 'From all that I hear from several quarters, it seems that Oxford did the subject great good. It is of enormous importance, the showing the world that a few first-rate men are not afraid of expressing their opinion.'[1] To Huxley Darwin confessed that he would as soon have died as tried to answer the Bishop in such an assembly—which may well be true—but Huxley's triumph has come to epitomize more than polarization of opinion, nothing less in fact than the victory of science over religion. When A. D. White (cf. Unit 1) wrote his *A History of the Warfare of Science with Theology*, he could describe the attempts to denigrate the Darwinian theory as the final effort of theology in a battle she was destined to lose.

Figure 14 Charles Kingsley, engraved by D. J. Pound from a photograph by Mayall (Mansell Collection).

But how accurate a description is it? Not all who opposed Darwin did so for reasons of religious prejudice; nor were all clergymen hostile. As early as February 1860 Darwin could inform the distinguished American botanist Asa Gray that 'several clergymen go far with me'—one of whom was the Reverend Charles Kingsley. And when Gray himself published a pamphlet with the title *Natural Selection Not Inconsistent With Natural Theology*, Darwin considered it of great service in quelling opposition and in making converts to Evolution. This surely means that we must look more closely at some of the less naïve responses to the Darwinian theory. Had it really undermined the theological enterprise? Could there still be arguments for design if not from design?

10 The Theological View of the Question

Figure 15 Asa Gray (Bettmann Archive).

There is no doubt that Darwin found what he called the theological view of the question painful. 'This is always painful to me', he informed Asa Gray. 'I am bewildered. I had no intention to write atheistically. But I own that I cannot see as plainly as others do, and as I should do, evidence of design and beneficence on all sides of us.' Darwin's own dilemma was partly occasioned by the fact that he saw *more* plainly than others did the world of imbalance and carnage which his theory presupposed. Whereas others could with difficulty put on the spectacles of natural selection, and quickly take them off again, Darwin had lived with his theory sufficiently long to need no spectacles. 'I cannot persuade myself', he confided to Gray, 'that a beneficent and omnipotent God would have designedly created the Ichneumonidae with the express intention of their feeding within the living bodies of Caterpillars, or that a cat should play with mice. Not believing this I see no necessity in the belief that the eye was expressly designed.' The fact is that a great effort of the imagination was required for most of Darwin's contemporaries to see the world as Darwin himself saw it. There was almost a psychological block to a full understanding of the theory. One could comprehend it at the rational level, and yet find oneself resisting the image of nature which Darwin had portrayed. It was true as Darwin had insisted, that to overcome the difficulties of the theory, one's reason had to conquer one's imagination, but it was no less true that to 'get inside' the theory, as it were, one's imagination had to overcome one's imagination.

This may sound like a play on words, but only to those whose imagination has failed them. Darwin well knew what was involved, for he had experienced it himself. 'If you are *in even so slight a degree* staggered,' he wrote to his old friend Henslow, 'then I am convinced with further reflection you will become more and more staggered, for this has been the process through which my mind has gone.'[2] One had to be staggered to understand, and there was no true convert who was not. 'Dr. Hooker,

[1] F. Darwin (ed.), *op. cit.*, vol. II, pp. 323–4.
[2] F. Darwin (ed.), *op. cit.*, vol. II, p. 218.

whom I consider one of the best judges in Europe, is a complete convert, and he thinks Lyell is likewise; certainly, judging from Lyell's letters to me on the subject he is deeply staggered.'[1] The psychological block had to be removed, and it was not easy. And even to be staggered was no guarantee that one would *embrace* the theory, as the case of Lyell would demonstrate! Can you suggest why the block was so great? In the *Origin* itself, Darwin had written this:

> Nothing is easier than to admit in words the truth of the universal struggle for life, or more difficult—at least I have found it so—than constantly to bear this conclusion in mind. Yet unless it be thoroughly engrained in the mind, I am convinced that the whole economy of nature, with every fact on distribution, rarity, abundance, extinction and variation, will be dimly seen or quite misunderstood.

The reason for stressing this point is a simple one. Among those who felt able to incorporate the idea of man's evolution within their religious, or indeed social and political, systems were many who had not been staggered, and who consequently came up with rather facile solutions. This is not the place for a systematic account of the religious response to Darwin, or even for a discussion of the relationship between the Darwinian controversy and the internal debates within the Anglican Church concerning the propriety and methods of Biblical criticism. These are themes to be developed by Dr Richardson in your next unit.

As a preparation for Unit 14, however, you might find it instructive to examine the series of religious responses set out below. Some are favourable, others less so, but your aim should be to decide whether each one does or does not do justice to Darwin's theory, and if it does not, why not. You should then answer the questions which follow the extracts.

A

'Let us rather look with calmness, even with hope and good-will, on these new theories; they surely mark a tendency toward a more, not a less, Scriptural view of Nature.

'Of old it was said by Him, without whom nothing is made. "My Father worketh hitherto, and I work." Shall we quarrel with Science if she should show how these words are true? What, in one word, should we have to say but this: "We know of old that God was so wise that he could make all things; but, behold, he is so much wiser than even that, that he can make all things make themselves." '

B

Wherefore, so long as gradatory, orderly, and adapted forms in Nature argue design, and at least while the physical cause of variation is utterly unknown and mysterious, we should advise Mr Darwin to assume, in the philosophy of his hypothesis, that variation has been led along certain beneficial lines. Streams flowing over a sloping plain by gravitation (here the counterpart of natural selection) may have worn their actual channels as they flowed; yet their particular courses may have been assigned; and where we see them forming definite and useful lines of irrigation, after a manner unaccountable on the laws of gravitation and dynamics, we should believe that the distribution was designed.

C

I grieve to say that I cannot honestly go as far as you do about Design. I am conscious that I am in an utterly hopeless muddle. I cannot think that the world, as we see it, is the result of chance; and yet I cannot look at each separate thing as the result of Design. To take a crucial example, you lead me to infer that you believe 'that variation has been led along certain beneficial lines.' I cannot believe this; and I think you would have to believe that the tail of the Fantail was led to vary in the number and direction of its feathers in order to gratify the caprice of a few men.

D

Darwinian teleology has the special advantage of accounting for the imperfections and failures as well as for successes. It not only accounts for them, but turns them to practical account. It explains the seeming waste as being part and parcel of a great economical

[1] F. Darwin (ed.), *op. cit.*, vol. II, p. 222.

process. Without the competing multitude, no struggle for life; and without this, no natural selection and survival of the fittest, no continuous adaptation to changing surroundings, no diversifications, and improvement, leading from lower up to higher and nobler forms. So the most puzzling things of all to the old-school teleologists are the *principia* of the Darwinian. In this system the forms and species, in all their variety, are not mere ends in themselves, but the whole a series of means and ends, in the contemplation of which we may obtain higher and more comprehensive, and perhaps worthier, as well as more consistent, views of design in Nature than heretofore. At least, it would appear that in Darwinian evolution we may have a theory that accords with if it does not explain the principal facts, and a teleology that is free from the common objections.

E

The action which [Darwin] attributes to natural selection is clearly regulated action. Why should natural selection favour the preservation of useful varieties only? Such action cannot be referred to blind force; it can belong to mind alone.

F

In saying that this system is atheistic, it is not said that Mr Darwin is an atheist. He expressly acknowledges the existence of God; and seems to feel the necessity of his existence to account for the origin of life. Nor is it meant that everyone who adopts the theory does it in an atheistic sense. . . . Natural laws are said to be to God what the chisel and the brush are to the artist. Then God is as much the author of species as the sculptor or painter is the author of the product of his skill. This is a theistic doctrine. That, however, is not Darwin's doctrine. His theory is that hundreds or thousands of millions of years ago God called a living germ, or living germs, into existence, and that since that time God has no more to do with the universe than if He did not exist. This is atheism to all intents and purposes.

G

Amid all the millions of mechanisms and instincts in the animal kingdom, there is no one instance of a mechanism or instinct occurring in one species for the exclusive benefit of another species, although there are a few cases in which a mechanism or instinct of benefit to its possessor has come to be utilized by other species. Now, on the beneficent design theory it is impossible to explain why, when all the mechanisms in the same species are invariably correlated for the benefit of that species, there should never be any such correlation between mechanisms in different species, or why the same remark should apply to instincts. For how magnificent a display of Divine beneficence would organic nature have afforded if all, or even some, species had been so inter-related as to minister to each other's necessities.

H

That the Struggle for Life has been a prominent actor in the drama is certain. Further research has only deepened the impression of the magnitude and universality of this great and far-reaching law. But that it is the sole or even the main agent in the process of Evolution must be denied. Creation is a drama, and no drama was ever put upon the stage with only one actor. The Struggle for Life is the 'Villain' of the piece, no more; and, like the 'Villain' in the play, its chief function is to re-act upon the other players for higher ends. There is, in point of fact, a second factor which one might venture to call the *Struggle for the Life of Others*, which plays an equally prominent part. . . . Sympathy, tenderness, unselfishness, and the long list of virtues which make up Altruism, are the direct outcome and essential accompaniment of the reproductive process. Without some rudimentary maternal solicitude for the egg in the humblest forms of life, or for the young among higher forms, the living world would not only suffer, but would cease. For a time in the life history of every higher animal the direct, personal, gratuitous, unrewarded help of another creature is a condition of existence.

1 The author of one of these passages is Darwin himself. Can you work out which one it is most likely to be?

2 One author acknowledges his confusion. Why was he confused?

3 Two or three of the passages reveal a possible misunderstanding of Darwin's theory. Which are they? How would you justify your choice?

4 The author of D claims that a Darwinian teleology has an advantage over more traditional versions of the argument from design. What is it supposed to be?

5 Both B and D were written by the same man, but is his contention in B a necessary premiss for his argument in D to work?

6 The author of one passage was trying to establish that from a conscientious examination of nature one would have to concede that if there were a Mind behind the evolutionary process it appeared not to

be 'of a nature which in man we should call moral'. Which extract best fits this description, and what is the argument?

---

1 C. You might have opted for either C or G, though the reference to the pigeon in C should have given the game away!

2 Darwin himself admits that he is confused: his own theory has no need of the concept of design to explain the details of adaptation, but he found it hard to believe that the world in its entirety was the product of chance.

3 I would choose B, E and H. B because Darwin (in C) takes exception to the notion of variations having been *led* in favourable directions; E because what Darwin *meant* by natural selection was the preservation of useful varieties and the extermination of the less well adapted; and H, because the author, Henry Drummond,[1] not only minimizes the agency of natural selection, but adds a new concept of his own—the struggle for the life of others—which, though it might be applicable in the case of man, was surely, in the animal world, referring to an instinct of the kind which Darwin had already explained, in principle at least, by natural selection.

4 The contention of Asa Gray[2] is that whereas the more traditional versions of the argument from design were open to objections based on wastage, and, by implication, pain and suffering, this is not the case with teleological arguments based on evolution, since Darwin has shown that *apparent* wastage in nature is an indispensable part of the process. Without it man would never have come into existence at all.

5 I would have thought not. The argument in D seems to presuppose a God in some measure bound by his own evolving creation, whereas the notion of His *leading* variations along choice lines might be open to the objection that if he could so lead why was there need for 'wastage' at all? (You might like to know that Gray's argument in D is taken from a later essay than B, by which time he was refraining from talk about variations having been led. He still insisted, however, that variation 'is not a product of, but a response to, the action of the environment'.)

6 G would be the best answer here. Its author, George Romanes,[3] having wrestled for many years with the logical problems raised by Darwin's theory, and having during that time followed a very personal religious pilgrimage, embracing and renouncing theism, embracing and renouncing a complete scepticism, argued that it was not possible to infer the existence of a *moral* Being from the evolutionary process. In order to evince *beneficent* design, nature would have to show a very different face—show adaptations which might be of exclusive benefit to other species.

It should be clear to you now that intelligent men could respond in very different ways to that crucial question: could natural selection be compatible with natural theology? Some, like Charles Hodge of the Princeton theological seminary, and author of paragraph F above, were adamant that Darwin's theory was effectively atheistic if not intentionally so.[4] Others, like Charles Kingsley, author of paragraph A, could convince themselves that Darwin had given them a loftier view of creation.[5] Romanes could believe in a Mind behind the process, but one whose moral nature, if indeed it made sense to describe it as such, could not be inferred from the facts of nature. The scope for a scheme of designed evolution was clearly there, although some apologetic strategies were decidedly ephemeral. The suggestion that variations were specially directed, though an early ploy of Asa Gray and of the popular writer, the Duke of Argyll, was soon open to the objection that here was just another case of the 'God-of-the-gaps'—a deity who could be ejected once the causes of variation were discovered. Much the same objection applied to the familiar plea that Darwin had not even pretended to explain the origin of life.

[1] H. Drummond, *The Ascent of Man*, London, 1904 edition, pp. 16–23.
[2] A. Gray, *Darwiniana*, edited by A. Hunter Depree, Harvard University Press, 1963, pp. 310–11.
[3] G. J. Romanes, *Thoughts on Religion*, 6th edition, London, 1896, p. 84.
[4] J. Dillenberger, *Protestant Thought and Natural Science*, Collins, 1961, pp. 242–3.
[5] C. Kingsley, as cited by A. Gray, *op. cit.*, p. 232.

Figure 16 T. H. Huxley (by kind permission of the President and Council of the Royal College of Surgeons).

A more effective strategy was to argue that the trail of evolution was not quite as tormented and bloodstained as Darwin's own canvas depicted. One could still be staggered and yet disagree with Huxley when he affirmed 'the moral indifference of nature' and 'the unfathomable injustice of the nature of things'. No less an authority than Wallace took issue with Huxley on precisely this point. Were our ears sharp enough, Huxley had concluded, we should hear incessant sighs and groans of pain like those heard by Dante at the gate of hell. 'Greatly exaggerated', was Wallace's reply. 'There is . . . good reason to believe . . . that the supposed "torments" and "miseries" of animals have little real existence, but are the reflection of the imagined sensations of cultivated men and women in similar circumstances; and that the amount of actual suffering caused by the struggle for existence among animals is altogether insignificant.'[1] That Wallace could so criticize Huxley provides a good illustration of the point we shall develop in the next section—that when it came to the interpretation of Darwin's universe, prior convictions played a decisive role. The eye of faith would, in Kingsley's words, see God's making things make themselves, the eye of agnosticism would see things making themselves. Whereas Asa Gray boldly incorporated the wastage of nature within a wider teleology (cf. D above), Darwin himself could not—at least not with conviction. Gray's argument that waste and suffering have been *shown* to be necessary for a creative evolutionary process to work, was a gift to the theologians wrestling with the problem of suffering—a gift which they were, perhaps, slow to appreciate—but a gift which many could not accept because the name of Darwin was associated with an agnosticism to which they would make no concession. 'I feel most deeply', Darwin had confided to Gray, 'that the whole subject is too profound for the human intellect. A dog might as well speculate on the mind of Newton. Let each man hope and believe what he can.'[2]

11 The Success of the Theory

To do justice to all the considerations which help to explain the rapid success of Darwin's ideas would require a book in itself; but, in this last section of the unit, we can at least introduce some of them. In the first place, of course, the theory was rapidly disseminated among men of science because, as Huxley insisted, Darwin had given them a naturalistic explanation for the origin of species which could at last be

[1] A. R. Wallace, *Darwinism*, London, 1905, p. 37.
[2] F. Darwin (ed.), *op. cit*., vol. II, p. 312.

taken seriously. Despite Darwin's own rethinking on the sufficiency of natural selection and despite protracted discussion on the significance of saltations, Darwin had explained too much for his theory to be jettisoned. In Section 6 you were asked to tabulate the facts which Darwin considered to be explained more successfully by his theory than by one of separate creation. As you have just seen, Darwin's antithesis did not rule out the possibility of creation by evolution, but the facts in favour of evolution itself were impressive indeed. In Huxley's account of the reception of the *Origin* which you were invited to read in Section 8, he could say that within the space of twenty-five years the 'struggle for existence' and 'natural selection' had become household words. Even those who had pitted their scientific knowledge against Darwin, had to concede that his ideas had been assimilated far more effectively than they could have wished. In John Tyndall's Belfast address to the British Association for the Advancement of Science (August 1874), there is a sensitive reminiscence of a meeting near Boston at which both he and Agassiz had been present—Agassiz having fought the evolutionists to the last:

> Rising from luncheon, we all halted as if by common consent, in front of a window, and continued there a discussion which had been started at table. The maple was in its autumn glory, and the exquisite beauty of the scene outside seemed, in my case, to interpenetrate without disturbance the intellectual action. Earnestly, almost sadly, Agassiz turned, and said to the gentlemen standing round, 'I confess that I was not prepared to see this theory received as it has been by the best intellects of our time. Its success is greater than I could have thought possible.'[1]

Its success back in Britain owed much to the energetic activity of Huxley who, when given the opportunity to address the public in lectures and 'lay sermons', always advertised the *scientific* merits of the Darwinian case, and one merit above all. Only Darwin's theory, Huxley insisted, could explain the palaeontological fact that certain animal types had persisted through a great time span, with little change at all compared with everything about them. 'It is one remarkable peculiarity of Mr Darwin's hypothesis', Huxley would claim, 'that it involves no necessary progression or incessant modification, and that it is perfectly consistent with the persistence for any length of time of a given primitive stock, contemporaneously with its modifications.' The most famous example of this is perhaps the coelacanth which is still at large today.

From your knowledge of the hypotheses of Lamarck, Chambers and Darwin, can you state how Huxley would have justified his proposition?

Lamarck, with his 'internal' mechanism for development, and Chambers, with his inexorable law of development, would presumably be more embarrassed by the non-evolution of certain types like the coelacanth. For the disciple of Darwin this was not a problem since, as Huxley himself explained, 'a primitive stock and its modifications may, occasionally, all find the conditions fitted for their existence; and though they come into competition, to a certain extent, with one another, the derivative species may not necessarily extirpate the primitive one, or *vice versa*'.[2] An organism perfectly adapted to a stable environment would not necessarily be subject to the pressures which would encourage the accumulation of further variations.

The scientific merits of Darwin's theory could be readily appreciated, and its own adaptability—its being sufficiently programmatic and versatile to accommodate all the facts of palaeontology—stood it in good stead. But the name of Darwin became a household word for other reasons too. One did not have to be very bright to recognize that Darwin had something interesting to say about 'the proper study of mankind':

[1] J. Tyndall, *Fragments of Science*, 6th edition, London, 1879, vol. II, p. 182.
[1] T. H. Huxley, *Lectures and Essays*, Cassell, pp. 106–7.

man. The public debate concerning *man's* place in nature spread the evolutionists' doctrine far and wide. Bent on reaching the 'wider circle of the intelligent public' Huxley prepared himself for the fray by reading all he could on the man-like apes, and by comparing human skulls and skeletons with those of the baboon, gorilla, and that pathetic subject of eighteenth-century experiments in elocution, the orang-outang. 'Even in the important matter of cranial capacity,' he would write, 'men differ more widely from one another than they do from the Apes.' And for the skull, no less than for the skeleton in general, 'the proposition holds good, that the differences between Man and the Gorilla are of smaller value than those between the Gorilla and some other Apes'.[1] With unrelenting zeal the orang's foot was shown to differ more widely from the foot of the gorilla than the latter did from that of man. With great determination, the celebrated anatomist Richard Owen tried to find a lobe in the human brain that could be found nowhere else, but it was Huxley's strategy that prevailed: 'if Man be separated by no greater structural barrier from the brutes than they are from one another—then it seems to follow that if any process of physical causation can be discovered by which the genera and families of ordinary mammals have been produced, that process of causation is amply sufficient to account for the origin of Man.'

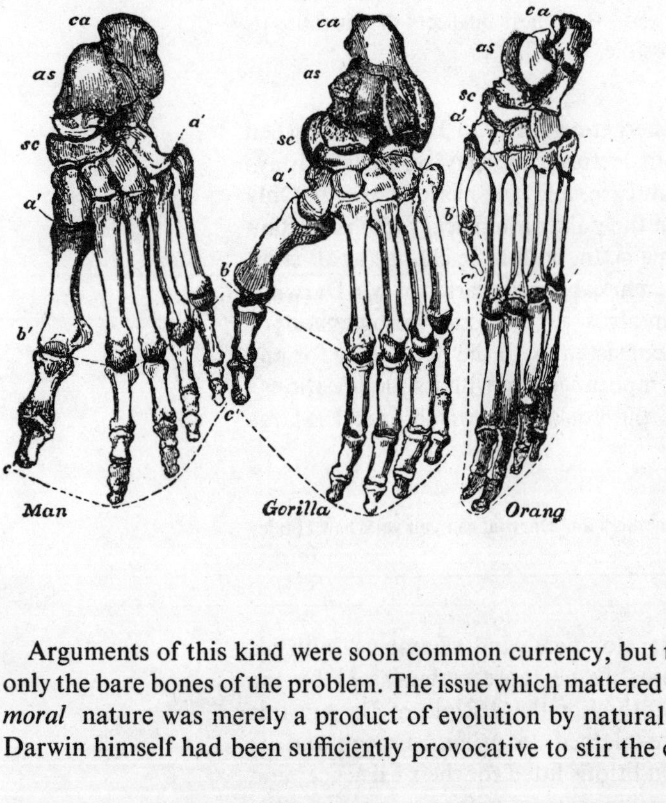

Figure 17 Foot of a man, a gorilla and an orang-outang of the same absolute length, to show the differences in proportion of each. Reduced from original drawings by Mr Waterhouse Hawkins (from T. H. Huxley, Letters and Essays. *Courtesy of Cassell).*

Arguments of this kind were soon common currency, but they did of course touch only the bare bones of the problem. The issue which mattered most was whether man's *moral* nature was merely a product of evolution by natural selection. On this point Darwin himself had been sufficiently provocative to stir the conscience:

> Ultimately a highly complex sentiment, having its first origin in the social instincts, largely guided by the approbation of our fellow-men, ruled by reason, self-interest, and in later times by deep religious feelings, confirmed by instruction and habit, all combined, constitute our moral sense or conscience.[2]

To those who still thought of their conscience as a divine 'still, small voice', this must have come as a hard saying, and one which was hotly disputed. The problems involved were too difficult even for a well-educated public, but it became virtually impossible to disregard them altogether—especially when it transpired that Wallace

[1] Huxley, *op. cit.*, p. 265.
[2] Cited by W. Coleman, *Biology in the Nineteenth Century*, Wiley, 1971, p. 85.

was now parting company with Darwin. There were some attributes of man which, according to Wallace, must have had a distinctly spiritual origin. This was his argument, and you should study it carefully:

> It appears then, that, both on account of the limited number of persons gifted with the mathematical, the artistic, or the musical faculty, as well as from the enormous variations in its development, these mental powers differ widely from those which are essential to man, and are, for the most part, common to him and the lower animals; and that they could not, therefore, possibly have been developed in him by means of the law of natural selection.[1]

Can you explain what Wallace was driving at here? Why would he feel entitled to use the word 'therefore' as he does?

He was driving at a distinction between mental powers that are somehow essential to man, and others such as musical talent which are not. The fact that mathematical, musical or artistic prowess is so rare suggests that these powers have not been necessary or even useful for survival. Accordingly they could not have been developed by a process of natural selection. They were, for Wallace, evidence of man's spiritual nature.

This was not an argument likely to convince a sceptic, but the very fact that the two pioneers of natural selection should appear to disagree on the ultimate nature of man was bound to attract attention and to add a particular flavour to the debate.

The scientific merits of the theory, and its controversial application to man, begin to explain its rapid percolation through society. But there are considerations of another kind which must be taken into account. Scientific ideas have often been accepted and assimilated because they could be made to serve some ulterior purpose. In earlier units we have seen how Newton's theories could be assimilated by Christian apologists such as Richard Bentley and made subservient to the needs of their natural theology. In the case of Darwin's theory with its emotive metaphors and far reaching implications for man's image of himself, the scope for exploitation was considerable. As the authority of Darwin was invoked to justify a whole spectrum of social, political and religious attitudes, so the prestige of 'Darwinism' grew. This is important because it illustrates the fact that the successful dissemination of scientific ideas may depend to no small extent on the state of society, and the tensions and prejudices within it. In the context of religious attitudes, Darwin recognized this for himself, not simply because he had been exposed to the abuse of clerical tongues, but because every man's judgement was so affected by his preconceptions that rational argument could hardly be the final arbiter. In a letter to Karl Marx he declared: 'Although I am a keen advocate of freedom of opinion on all questions, it seems to me (rightly or wrongly) that direct arguments against Christianity and Theism hardly have any effect on the public. . . .' Marx himself had claimed, as early as January 1861, that Darwin's book had furnished him with a scientific licence for the class struggle in history. Human preconceptions do act as filters in this way, and Darwin's concept of the struggle for existence was soon being used to justify political ideologies of every hue. In Unit 12 we stressed that from a strictly scientific standpoint the traditional 'precursors of Darwin' were scarcely precursors at all, but this does not mean that their theories had not been invoked by the philosophers of *social* evolution in the period before Darwin. Indeed it was Lyell's 'refutation' of Lamarck that appears to have played a decisive role in convincing Herbert Spencer that Lamarck's ideas on adaptation would serve very well for a doctrine of inexorable social progress.[2]

[1] Wallace, *op. cit.*, p. 472.

[2] R. Young, 'The Historiographic and Ideological Contexts of the Nineteenth-Century Debate on Man's Place in Nature', in M. Teich and R. Young (eds.), *Changing Perspectives in the History of Science*, Heinemann, 1973, p. 410.

Once Darwin's theory was published it could be used to reinforce and confer respectability upon schemes of social evolution that were already in existence. In other words Darwin's metaphors could become household words, precisely because they were so malleable when detached from their original biological context.

To help you to grasp the full force of this argument there are several extracts from nineteenth-century sources reproduced below. Each one provides a different clue to the manner in which Darwin's theory could be exploited for ulterior purposes. Study these extracts now with a view to deciding what the ulterior purpose could be in each case:

A

We allow everybody to have his own taste. We know that there are many who have more respect for a count or a baron gone to waste by his slovenliness than for a commoner who has reached the top by his talent and industry. Our taste is the reverse. We feel that humanity has much more reason to be proud of having gradually worked itself up by the continuous effort of innumerable generations from miserable, animal beginnings to its present status. We prefer this to being the descendants of a couple, who, having been created in the image of God, were kicked out of Paradise and to knowing that we are still far from having reached the level from which they fell in the beginning. Nothing dampens courage as much as the certainty that something we have trifled away can never be entirely regained. But nothing raises courage as much as facing a path, of which we do not know how far and how high it will lead us yet.

B

As surely as a blacksmith's arm grows large, so surely must the human faculties be moulded into complete fitness for the social state; so surely must the things we call evil and immorality disappear; so surely must man become perfect. . . . Evolution can end only in the establishment of the greatest perfection and the most complete happiness. . . . Progress . . . is not an accident, but a necessity.

C

Many economists seem to be terrified that distress and misery still remain on earth and promise to remain as long as the vices of human nature remain. Many of them are frightened at liberty, especially under the form of competition, which they elevate into a bugbear. They think it bears harshly on the weak. They do not perceive that here 'the strong' and 'the weak' are terms which admit of no definition unless they are made equivalent to the industrious and the idle, the frugal and the extravagant. They do not perceive, furthermore, that if we do not like the survival of the fittest, we have only one possible alternative, and that is the survival of the unfittest. The former is the law of civilization; the latter is the law of anti-civilization. We have our choice between the two, or we can go on, as in the past, vacillating between the two, but a third plan—the socialist desideratum—a plan for nourishing the unfittest and yet advancing in civilization, no man will ever find.

D

Then will the world enter upon a new stage of its history—the final competition of races for which the Anglo-Saxon is being schooled. If I do not read amiss, this powerful race will move down upon Mexico, down upon Central and South America, out upon the islands of the sea, over upon Africa and beyond. And can anyone doubt that the result of this competition will be the 'survival of the fittest'?

E

It is the mind thus stored with the choicest materials of the teleologist that rejects teleology, seeking to refer these wonders to natural causes. They illustrate, according to Darwin, the method of nature, not the 'technic' of a manlike Artificer. . . . Diminishing gradually the number of progenitors, Mr Darwin comes at length to one 'primordial form'; but he does not say, so far as I remember, how he supposes this form to have been introduced. He quotes with satisfaction the words of a celebrated author and divine who had 'gradually learnt to see that it was just as noble a conception of the Deity to believe He created a few original forms, capable of self-development into other and needful forms, as to believe He required a fresh act of creation to supply the voids caused by the action of His laws.' What Mr Darwin thinks of this view of the introduction of life, I do not know. But the anthropomorphism, which it seemed his object to set aside, is as firmly associated with the creation of a few forms as with the creation of a multitude. We need clearness and thoroughness here. Two courses and two only are possible. Either let us open our doors freely to the conception of creative acts, or abandoning them, let us radically change our notions of Matter. . . . Divorced from matter, where is life? Whatever our *faith* may say, our *knowledge* shows them to be indissolubly joined. Every meal we eat, and every cup we drink, illustrates the mysterious control of Mind by Matter.

F

Mr Mivart does not hesitate to push his attempt to harmonize science with Catholic orthodoxy to its utmost limit; and, while assuming that the soul of man 'arises from immediate and direct creation', he supposes that his body was 'formed at first (as now in each separate individual) by derivative, or secondary creation, through natural laws'. . . . [But], if Suarez has rightly stated Catholic doctrine, then is evolution utter heresy. And such I believe it to be. In addition to the truth of the doctrine of evolution, indeed, one of its great merits in my eyes, is the fact that it occupies a position of complete and irreconcilable antagonism to that vigorous and consistent enemy of the highest intellectual, moral, and social life of mankind—the Catholic Church. No doubt, Mr Mivart, like other putters of new wine into old bottles, is actuated by motives which are worthy of respect, and even of sympathy; but his attempt has met with the fate which the Scripture prophesies for all such.

What ulterior purpose can you detect, and how is Darwin's theory appropriated in each case?

A In this passage the German philosopher and theologian D. F. Strauss[1] is using Darwin's theory to justify a more optimistic theology of man than would be allowed by traditional perspectives based on the doctrine of the Fall. The name of Darwin could be invoked both to cast doubt on the historicity of the creation story in Genesis, and to sanction the view that man had risen, not fallen.

B The extracts from Spencer[2] indicate that the idea of a progressive adaptation to environment could be used to justify an ethical theory which could look upon evil and immorality not as essential to the nature of things, but as a very transient inconvenience. Once man becomes fully adapted to his environment all vice will disappear. Though Spencer's theory had its roots in Lamarck's development hypothesis, it was certainly possible to appeal to Darwin's authority in order to countenance the view than man's selfishness was merely a vestige of his animal past. A belief in the inevitability of progress could be read into Darwin's remark that natural selection invariably worked for the improvement of each species.

C The passage from Sumner,[3] who was one of Spencer's most ardent American disciples, illustrates the way in which 'the survival of the fittest' could be put on parade as an anti-socialist slogan. If the fittest automatically survive, then any political interference must violate nature's law, and so lead to the encouragement and preservation of weakness. In this way the name of Darwin could be cited to justify a Protestant ethic, a capitalist society, or any laissez-faire economy.

D That Darwin's theory could be invoked to support a racist creed is clear from this vision of the Reverend Josiah Strong.[4] If the struggle for existence was translated into the language of racial competition, then the vigour and success of the Anglo-Saxon race could be used to demonstrate its intrinsic superiority. Variants of this argument were used to countenance imperialism, and, nearer to our time, the anti-semitism of Hitler.

E The excerpts from John Tyndall's *Belfast Address*[5] show how Darwin's theory could come to the aid of the 'materialist'. Once the need to postulate acts of creation was eliminated, the door was wide open for a philosophy which, by attributing new powers to matter itself, could stress the subordination of mind to matter.

F Huxley's pugnacious attitude towards Mivart, who had managed to reconcile

[1] Cited by E. Benz, *Evolution and Christian Hope*, Gollancz, 1967, p. 74.

[2] J. Passmore, *The Perfectibility of Man*, Duckworth, 1970, p. 241.

[3] W. Sumner, as cited by R. Hofstadter, *Social Darwinism in American Thought*, Beacon Press, Boston, 1955, p. 57.

[4] Hofstadter, *op. cit.*, p. 179.

[5] Tyndall, *op. cit.*, pp. 179, 191–2.

Catholicism and evolution, illustrates the way in which the name of Darwin was soon associated with anti-Christian propaganda, and anti-Catholicism in particular.[1]

You should by now have grasped the point that in the dissemination of Darwin's theory, ulterior motives of various kinds could have a large part to play. The irony is that Darwin himself would have dissociated himself from most if not all of these sentiments. He did feel passionately on certain political issues, and early in life had conceived a hatred of slavery. But in elaborating his theory it had not occurred to him that an explanation for the origin of species could be relevant to the social problems of mankind. He might have shared Strauss' belief that the ascent of man was both ennobling and edifying, but he was perhaps too much of a 'Liberal or a Radical' to acquiesce in Spencer's optimism, and too humane to pit the strong against the weak as Sumner did. Though the subtitle of the *Origin* was *The Preservation of Favoured Races in the Struggle for Life*, Darwin would have been horrified by the militant racism with which his name was later to be associated. Josiah Strong could quote a passage from *The Descent of Man* in favour of the unusual courage and energy of the American people, but Darwin's intention had simply been to illustrate the action of natural selection on Anglo-Saxon emigrees. Nor is it likely that Darwin would have condoned the materialism of Tyndall. That matter had been much maligned was not one of his grievances, and the most sensitive of his readers recognized the fact: 'No doubt Mr Darwin cannot be charged with materialism', wrote Richard Holt Hutton in an essay in 1874. 'His writings, though they seem to throw doubt on the conception which makes the minutest phenomena of Nature the expression of a Divine purpose, and harp a good deal on the failures of the organic world, seem to me full of belief in an ultimate intellectual origin for the scheme of things.'[2]

And finally, how could a man who had written, 'Let each man hope and believe what he can', have approved the manner in which Huxley was molesting Mivart? Darwin himself had gradually relinquished the Christian faith during the 1830s, and that may help to explain why he was unable to perceive the design in nature that Asa Gray assured him was still there to be seen. The motive force behind the *desire* to see was no longer there. But he could never bring himself to take up the offensive against the faith which was still so dear to Emma his wife. We tend to think of science versus religion in Victorian society, but Darwin's own crisis of faith, if it can be called a crisis, serves only to show how simplistic a picture that must be. Certainly his scientific discoveries would have reinforced his growing scepticism, but one of his greatest difficulties with the Christian faith was not that there was no slot for Adam in the evolution of man but, according to his intimate friend Romanes, that so small a proportion of mankind had ever heard of Christ.[3]

[1] T. H. Huxley, *Darwiniana*, London, 1893, pp. 142, 147.

[2] R. H. Hutton, *Aspects of Religious and Scientific Thought*, London, 1899, Gregg International Reprint, p. 47.

[3] Romanes, *op. cit.*, p. 182.

Further Reading

Nora Barlow (ed.), *The Autobiography of Charles Darwin*, Collins, 1958.
Sir Gavin de Beer, *Charles Darwin*, Thomas Nelson and Sons Ltd., 1963.
Francis Darwin (ed.), *The Life and Letters of Charles Darwin*, 3 volumes, London, 1887.
Alan Moorehead, *Darwin and the Beagle*, Penguin, 1971.
P. J. Vorzimmer, *Charles Darwin: The Years of Controversy*, University of London Press, 1972.
A. Ellegard, *Darwin and the General Reader*, Göteberg, 1958.
D. Daiches Raphael, 'Darwinism and Ethics', in S. A. Barnett (ed.), *A Century of Darwin*, Mercury Books, 1962, pp. 334–59.

Acknowledgements

It is a pleasure to acknowledge the advice of Dr Jonathan Hodge and members of the Course Team.

Grateful acknowledgement is made to the following sources for illustrations used in this unit:

Figure 1: William Collins Ltd; *Figures 2, 12, 13 and 14*: Mansell Collection; *Figures 3 and 4*: George Rainbird Ltd; *Figure 5*: David Lack and Scientific American; *Figure 6*: Courtesy of the Trustees of the British Museum (Natural History); *Figures 7, 9 and 16*: By kind permission of the President and Council of the Royal College of Surgeons; *Figure 8*: Courtesy of the Racing Pigeon Pictorial; *Figure 10*: Yale University Press; *Figure 11*: Miss J. Gosse and William Heinemann and Co.Ltd; *Figure 15*: Bettmann Archive; *Figure 17*: Cassell and Co. Ltd.

Unit 14
Religious Thought and the Idea of Evolution

Prepared by the Very Rev Dr Alan Richardson KBE for the Course Team

ILLUSTRATION OVERLEAF

'Nature red in tooth and claw': mid-nineteenth-century engraving by Webb of tiger and crocodile (Mary Evans Picture Library).

Contents

1 Introduction: the History of an Idea, 'Evolution'

The following are the main objectives of this unit, at the end of which you should be able to:

1 Place the concept of Darwinian evolution in the context of theological thought in the nineteenth century.
2 Assess the importance of Greek and Hebrew attitudes to history.
3 Identify the main causes towards changed approaches to the Bible in the nineteenth century.
4 Assess the importance of such changes for scientific thinking in general.
5 Identify the chief attitudes in the Victorian church to problems raised by Darwinism.

The only way in which we can understand an idea or an institution is by understanding its history, how in fact it came to be what it is today. Curiously enough, this truth was not properly appreciated until about two hundred years ago, as we shall see. It is strikingly illustrated by the study of the idea of evolution itself. Why did the concept about the evolution of the world, of man and of his ideas not become clear to the human mind until the nineteenth century A.D., and then only within a comparatively small section of 'western' civilization? We are compelled to look for an answer to this question by studying the evolution of the *idea* of evolution, because there is no other way of understanding any human enterprise or idea than by studying how it came to be what it is. It would be a good exercise, when you have studied the whole of this section, to write down in your own words what you think about the above statements. Is it true that you yourself have come to understand better what is meant by the idea of evolution through your study of how our present-day understanding of it came to be what it is? And then, perhaps, you might go on to ask yourself whether it is likely that the concept of evolution will in the twenty-first and twenty-second centuries be significantly different from the one which is generally held today. (There is much profit to be gained from asking ourselves such unanswerable questions, because the exercise inhibits any conceit we may have with regard to the inadequacies of our predecessors' conceptions.) We are all, scientists and non-scientists alike, creatures of our own time-conditioning. The following is the penultimate paragraph of Darwin's *Origin of Species*; please read and then answer the question that follows.

> We may look with some confidence to a secure future . . . And as natural selection works solely by and for the good of each being, all corporeal and mental endowments will tend to progress towards perfection.

How far was this an 'objective' scientific judgement or how far was it a spin-off from the prevailing Victorian attitudes? Could any sensitive scientist have reached such a conclusion in 1959?

In no obvious sense could this remark be said to reflect objective scientific judgement. It was, however, very much in accordance with the Victorian belief in progress, a belief that was nurtured by the rising standards of living enjoyed by the middle classes. No doubt the view would not have been shared by the less fortunate factory workers. Darwin's remarks are a classic expression of his *Weltanschauung* (or world view) which, while it must have affected his science, is no part of it. It is highly doubtful if a sensitive scientist in our own postwar era could commit himself to such a platitude. However, progress in the sense of better adaptation was a more 'scientific' concept (Unit 13).

There is a sense in which it could be said that the publication of Darwin's *Origin of Species* in 1859 marked the final disintegration of the mediaeval world view. The great cultural achievement of the high civilization of the Middle Ages resulted from its fusion of the two quite disparate currents of Greek philosophy and Hebrew religion in the articulated synthesis known as the Scholastic system. The paradoxical character of this improbable union of opposites has been strikingly expressed by Professor A. O. Lovejoy:

> The God of Aristotle had almost nothing in common with the God of the Sermon on the Mount—though by one of the strangest and most momentous paradoxes of Western history the philosophical theology of Christendom identified them, and defined the chief end of man as the imitation of both.[1]

Lovejoy was one of the 'founding fathers' of the study of the history of ideas and this quotation from his famous book is frequently cited.

How would you re-word the truth to which Lovejoy is calling attention?

To those who have studied the synthesis of Biblical and classical ways of thinking in the first four centuries A.D., the confluence, though 'momentous', is not so paradoxical or strange as it is thus made out to be. It has indeed an aspect of inevitability, as all significant historical developments appear in retrospect to have. So perhaps one should tone down the element of paradox, while at the same time fully retaining and emphasizing the importance of the Scholastic synthesis. One of the problems arising from this union is discussed in the following section.[2]

---

2 Greek and Hebrew Attitudes

Once again we need to be aware of the twofold heritage from antiquity. Before proceeding further, please look back to pp. 56–7 of Unit 2 where Professor Hooykaas deals with this matter with special reference to cosmology.

2.1 Greek and Hebrew ideas of history

Professor Hooykaas in *Religion and the Rise of Modern Science* has argued that 'the confrontation of Graeco-Roman culture with Biblical religion engendered, after centuries of tension, a new science' (p. 161). A similar conclusion could be reached about the way in which an entirely new understanding of history arose out of the confrontation. The Greek view was that history is cyclical; it repeats itself eternally. (As in astronomy, etc., the notion that the *circle* was the perfect Form or Idea dominated Greek thinking about history.) The world-historical process returned after a fixed number of years to the same point and then went round again: a circle has no beginning and no end. From all eternity the same Socrates would marry the same Xanthippe and drink the same fatal cup of hemlock as the sequence came round again. History

[1] A. O. Lovejoy, *The Great Chain of Being*, Oxford University Press, 1936, p. 5.

[2] For those who would perhaps some day wish to follow up the important development in the history of ideas to which Lovejoy calls attention, one might recommend Werner Jaeger, *Early Christianity and Greek Paideia*, Harvard University Press, 1965, and (at greater length) C. N. Cochrane, *Christianity and Classical Culture*, Oxford University Press, 1940.

was assimilated into nature; it was circular, like the movement of the heavenly bodies or the repetitive natural process of spring, summer, autumn and winter.[1]

How far is this view of history compatible with the Christian concept? (See, if necessary, *Hooykaas* (SET BOOK), p. 8.)

The Biblical view of history is linear, not circular. History has a beginning, a middle and an end (or goal). Hence in the Biblical view there can be purpose and progress in history. In the Greek view there can be neither purpose nor progress, because history is seen as a meaningless repetition of birth, growth, decay and death (the assimilation of history to the natural process of the seasons). If the Greek view of history had not been rejected in favour of the Hebraic view, it is less likely (though not entirely inconceivable) that a theory of evolution could have been developed, because the idea of evolution presupposes belief in real change, not merely a circular movement back to the starting-point: the conception of evolution in reverse would hardly carry credibility. For the Hebrews history was not absorbed into nature, as it was in Greek religion and in the nature-religions of the peoples of the Fertile Crescent from Egypt to Babylon. (The Hebrew language has no word meaning 'nature', the Greek *physis*.) Rather, God was seen as the Lord of Nature because He had *first* been perceived as the Lord of History.

An early example of the confrontation between the Greek and Judaeo-Christian ideas of history is recorded in Acts 17:16–34. How did Paul's reception illustrate the basic point of difference?

Here the antithesis is seen much more clearly than in Lovejoy's illustration of the Sermon on the Mount (where there is *some* common ground between the two ethical systems). Paul's rejection by the Greeks came with his reference to the Resurrection of Christ, a unique act of God in history that was totally at variance with the Greek conception.[2]

One of the truly momentous developments in the history of Western thought was the decisive rejection of the Greek cyclical conception of history by the theologians of the early Christian Church. Origen (*c.* 186–253) had protested against the notion that in another Athens another Socrates would marry another Xanthippe and drink another cup of hemlock (*Contra Celsum*, iv, 68). St Augustine (354–430) said that it would be stupid to believe that the same Plato had sat in the same Academy teaching the same pupils at fixed intervals through infinite ages in the past and would do so again through all the cycles of an infinite future, but it would be morally impossible to believe such nonsense about Christ—that another Pilate would crucify another Jesus on another cross to all eternity (*De Civitate Dei*, xii, 14). 'Christ being raised from the dead dieth no more; death hath no more dominion over him' (Romans 6: 9). This is the text which changed the outlook of European man on history. It is interesting to recall that Christian theologians had banished Greek notions about the cycle of history (and with them the associated notions of Fate, Necessity, Luck, Chance, etc.) more than a thousand years before the mediaeval and renaissance students of

[1] There were, of course, modifications to this view and some divergence as to the extent of the cycle. Some, for example, held that the 'new' Socrates would be similar to, but not identical with, the original one. Some applied the concept to the Universe as a whole while others (as Aristotle) restricted the cycle to human and local affairs. There is even in Empedocles and Anaximander some hint of an evolutionary conception. But on the whole the overwhelming impression is one of circularity, not progress.

[2] For further discussion of Paul's view of history see The Open University (1974) A291 *The Early Roman Empire and the Rise of Christianity*, Units 13–16 *The Rise of Christianity* (especially Section 9).

mechanics and astronomy had freed themselves from the animistic notions of Aristotelian physics or the conception of the Circle as the perfect Form of the motions of the heavenly bodies.

The view that historical process is repetitive has been held by the vast majority of the millions of people who have ever lived. Some of the great religions of the East, for example, held that the ultimate blessedness of the individual soul lay in its escape from the 'wheel of rebirth' into Nothingness (i.e. non-historical existence).

To what extent was this cyclical view of history banished from European thought in general?

While it is undoubtedly true that the patristic writings had showed the incompatibility between this view and the Biblical one the Greek ideas lingered on in popular thought for a dozen centuries and more. One has only to think of the hankerings after a Golden Age around 1700 to realize how persistent they were. See Kubrin's article in *Anthology II*, and television programme 7 on 'The Flood'. But as already noted they received a severe blow in the confrontation with Biblical religion in the Reformation and Renaissance.

Do you think that the scientific world view of today could have been achieved in a civilization which believed in the immutability of the repetitive cycle of world process? Can you suggest reasons for your view?

An important constituent in the scientific world view of today is something akin to the Darwinian theory of organic evolution, which could hardly have been conceived in any age which thought of history as essentially changeless, repetitive and (in this sense) static, and which maintained a philosophy that basically regarded progress as somehow illusory; perfection (the Real) was itself incapable of change, since any change would inevitably be an imperfection and therefore a logical impossibility.

2.2 The inspiration of Scripture

There was, however, one belief inherited by the Church from Greek thought, that was widely prevalent in the nineteenth century, although a few outstanding theologians (notably Luther) had begun to suspect its non-Hebraic origin at an earlier date.

This is the notion of 'inspiration' as it was taken over from Greek thought in the early Christian centuries. In point of fact the noun 'inspiration' is hardly a Biblical word at all; it is found only once in the Authorized Version (1611), at II Timothy 3:16: 'All Scripture is given by inspiration of God....' This translation was corrected in the Revised Version of 1881 and in all subsequent scholarly modern translations. Thus, the New English Bible reads 'Every inspired Scripture ...' The meaning against the background of Hebrew thought is that God has breathed life-giving truth into the Scriptures just as He breathed into man's nostrils the breath of life and man became a living soul (Genesis 2:7; cf. Wisdom 15:11 in the Apocrypha). But the Church in the age of the Greek Fathers assumed that this poetic (and very Hebraic) way of speaking meant that the Hebrew prophets and the Scriptural writers generally were 'inspired' after the manner of Vergil's Sibyl (*Aeneid*, vi) through whom the god literally spoke. Thus, Athenagoras in the second century represents the Holy Spirit as a flute-player who uses the prophet as his instrument. Hence a doctrine of literal verbal inspiration was gradually and almost unconsciously adopted into Christian thought. It remained the largely uncriticized view of the Church until the modern critical historical study of the Scriptures began to be developed towards the end of the eighteenth century, because there was no particular reason in any earlier period to call it into question.

In the post-Copernican Age how were scientific data commonly reconciled with Holy Scripture?

In the post-Copernican age it was not difficult for devoutly religious investigators like Kepler, who regarded his scientific discoveries as 'thinking God's thoughts after him', to conclude that Holy Scripture was not given to us for the purpose of conveying scientific information and that its language was accommodated to the capacities of ordinary people's ability to understand. This point of view remained the general attitude of the 'natural philosophers' (i.e. scientists) of the whole period from the sixteenth century to the first half of the nineteenth century, by which time it was agreed even amongst the clerical geologists (including Buckland at Oxford and Sedgwick at Cambridge) that the age of the earth was vastly greater than had been supposed. The six 'days' of creation in Genesis were only an example of how the Holy Spirit accommodated His revelation to the lowly understanding of mankind: e.g. the 'days' were not to be regarded as 24-hour sequences but were a way of speaking about the six stages in the age-long work of creation.

Except amongst scholars trained in the historical approach of the nineteenth century, the legacy of the Greek notion of 'inspiration', which had for so long been domiciled in Christian thought, was the chief source of the opposition to 'Darwinism' after 1859.

Can you suggest why this should have been so?

This could have happened only if the belief in 'inspiration' had carried with it certain other assumptions. One was that the words were literally dictated by God to scribes whose sole duty was to record them accurately, rather like a typewriter. The new-style Biblical scholars were well aware that the theory of the divine dictation of the actual words of Scripture could not have been derived from the study of the Biblical writings themselves. It was as clear to them as it is to us that St Paul, for example, did not think of himself as a mere 'pen in the hand of the Holy Spirit', for we can observe from his writing how he strove to think out the implications of the Gospel for himself (e.g. I Corinthians 7:10, 12, 25, 40).

This emphasis on verbal precision carried with it the further assumption that literal meaning was all that mattered. The astonishing thing is that such attitudes flew in the face not only of Biblical scholarship but also of the long-established view of 'accommodation'. We return to this point later. When Darwin's *Origin of Species* was published, only those clergy and laymen who had kept abreast of Biblical and historical scholarship, well developed by that date, were neither surprised nor alarmed by what was already an idea familiar to them. They recognized that Darwin had skilfully provided empirical evidence to corroborate the idea of real development in world process which had been articulated with increasing clarity and conviction by the new-style historians who had been at work for just about one hundred years before Darwin published the results of his biological researches in 1859. It is not surprising that those who were not familiar with the results of the Biblical and historical studies of the previous hundred years should have been shocked and grieved when it seemed as if the whole basis of Christian belief had been destroyed.

3 The Revolution in Historical Thinking

Scientific progress is achieved not so much by the making of new discoveries as by the attainment of new perspectives. Discoveries are made by looking at the well-known phenomena from a new angle, as when, for instance, Copernicus decided that it would be worthwhile to see how the mathematical situation would appear if one began with the assumption that the earth went round the sun. The justification of the new way of looking at the universe culminated in the publication of Newton's *Principia Mathematica* in 1687.

This very important achievement of looking at the world from the new *scientific* perspective did not, however, raise any radical questions concerning the relationship of science and religion—at least after the Copernican issue had been settled. The founders of the Royal Society (1662) were for the most part devoutly Christian men. In his *History of the Royal Society* (1702) the Rev Thomas Sprat was able to write: 'Of our churchmen, the greatest and most reverend, by their care and passion and endeavours in advancing this institution, have taken off the unjust scandal from natural knowledge, that it is an enemy to divinity. By the perpetual patronage and assistance they have afforded to the Royal Society, they have confuted the false opinions of those who believe that philosophers must needs be irreligious.'

3.1 The new historical perspective

No radically new *historical* perspective was to appear until about a century after the founding of the Royal Society. What was lacking was a sense of real development, whether of the modes of human existence or indeed of the universe itself. It is often said that Newton and his followers believed that the universe had come into existence all at once exactly as he saw it through his telescope, in the six days of creation of the Book of Genesis. There was no notion of either inorganic or organic evolution and no idea of development in human history. Things were just as they had always been.

In fact, so far was the seventeenth century from possessing any conception of development that it was widely assumed that human civilization had actually degenerated since the original creation of mankind. The Rev Robert South (1634–1716) declared that 'an Aristotle was but the rubbish of an Adam, and Athens but the rudiments of Paradise'. The degeneration had continued: it was not until the time of Dean Swift (cf. *The Battle of the Books*, 1704) that it was possible even to debate the question whether modern literature was worthy of comparison with the classics of antiquity. Even at the end of 'the Age of Reason' or 'the Enlightenment' (usually reckoned *c.* 1650–1780) it was still possible to see human history as 'decline and fall': cf. Gibbon's *Decline and Fall of the Roman Empire* (1776–88). In Gibbon's view 'the period in the history of the world during which the condition of the human race was most happy and prosperous' was 'that which elapsed from the death of Domitian to the accession of Commodus' (i.e. 96–192 A.D.), which includes the reign of Marcus Aurelius. The Middle Ages were the period of 'barbarism and religion'.

Thus, towards the end of the eighteenth century the greatest historical masterpiece in the English language was written from the perspective of the decline and fall of civilization, with hardly a hint of a renaissance.

Read the article by Kubrin in *Anthology II* (especially pp. 167–9) and suggest how Newton illustrates these two tendencies.

It appears that, for Newton, the universe was *not* conceived as wholly static, and that he viewed it rather as declining. But the additional point made by Kubrin is that this decline was often regarded in England as a matter for joy and millenial hope. Compare also the views of Burnet (television programmes 7 and 8). So it is not surprising that

Newton himself devoted more time to deciphering the secrets of the book of Revelation (and other apocalyptic literature) than he did to the study of the celestial bodies themselves.

Gibbon's pessimistic attitude, however, contrasted with the general tone of the *philosophes* of the 'Age of Reason', who had persuaded themselves that they were actually living in the period of history during which the condition of the human race was most happy and prosperous.[1] They spoke constantly of 'these enlightened times' and believed that they were already enjoying the amenities of life in the most developed civilization in the history of the human race, and one which could progress by the light of reason to ever higher levels of culture and happiness. This comfortable belief was based on the conviction that now at last, when the superstitions and evils of the age of 'barbarism and religion' had been banished by reason, the future happiness of the human race was assured: the heavenly city of the *philosophes* was a new Athens, Paris in the golden age which was now being ushered in. Within a generation or two the habit of reason, which would rule the world, would restore the glories of the age of Marcus Aurelius himself. As Monseigneur Knox wittily remarked, the *philosophes* 'who had lost the sense of religious certainty enrolled themselves under the banner of optimism; the world's future occupied their thoughts instead of a future world, and, by a kind of inverted Confucianism, they fell to worshipping their grandchildren'.

How far was the optimism of the *philosophes* justified, and to what extent did it relate to evolutionary thinking?

Their optimism proved ill-founded, since such of the *philosophes* as lived on until 1789 (notably Condorcet) did not survive the horrors of the Revolution. Moreover this optimism had nothing to do with any kind of doctrine of biological evolution; it was based solely on their belief that reason would shortly rule the world.

Already the geologists had pointed to facts which could be interpreted as evidence of evolution in prehistoric millenia, but the men of Enlightenment dismissed the matter airily. Voltaire (perhaps frivolously) suggested that the sea-shells found high up in the Alps had fallen from the hats of the pilgrims journeying to Rome. As you have seen elsewhere he was committed to deism which was certainly anti-evolutionary. Since the Enlightenment period belief in reason as an instrument of the ordering of human affairs has been undermined, chiefly by thinkers like Marx and Freud as well as by the course of events. Questions which concern men today relate to the prospects of reason as an instrument in the progress of mankind. Are there any alternatives to 'reason' in the *philosophes*' sense? Must we abandon rationalism altogether as a social gospel? Is there any sense in which we may dare to believe in 'progress' at all?

The *philosophes* wrote a great deal of history, but it was of a literary kind, undertaken not as research into the past but as propaganda for the present. But even before the end of the Age of Reason a second revolution in modern thought was beginning, quite as significant as the scientific revolution of the period from Copernicus to Newton. This was the revolution in historical method. The new history was scientific in the sense of being well researched, though it was not (as certain nineteenth-century positivists supposed it would become) scientific in the same way as physics and chemistry, i.e. it would not discover immutable laws or generalizations about human nature, unknown to the ancients, as the natural sciences were believed to have done in the case of the physical order. But it proceeded to demonstrate what Newton and the founders of the Royal Society had never suspected, namely, that the ancient world was not like the mediaeval or the modern world, but that on the contrary

[1] See also Unit 7, p. 37.

there had been a development, a process of real change, a progress rather than a decline, an advance from primitive beginnings to the 'enlightened' Age of Reason. A new breed of investigators, who can properly be called historians in our modern sense, was coming into being. Of course, before them there had been exceedingly learned experts on the past, but they would be more correctly described as antiquarians or archaeologists rather than historians. There was the solitary pioneer, the Neapolitan Vico (1668–1744), a lonely Melchizedeck who had no predecessors and no successors, of whom G. N. Clark remarked that his *Scienza Nuova* (1725) was 'not written until the eighteenth (century), not read until the nineteenth, and not understood until the twentieth'. Paul Hazard has said that the course of modern thought would have been very different if Vico had achieved a European reputation in his own age. Classical studies were highly developed, but there were no studies in history, as we know the subject today, in the universities of the eighteenth century and indeed until the second half of the nineteenth.

An interesting case is that of Richard Bentley (1662–1742). Displaying the skills of the later historians, he ranged far wider than history in his interests. Often said to be the greatest classical scholar England has ever bred, he not only produced the most acute exposure of a classical forgery (*The Letters of Phalaris*, 1699) but was actually appointed as the first Boyle lecturer (1692) as the person best qualified to defend the Christian faith on the basis of the Newtonian physics. Bentley disavowed any intention of arguing from the truth of the Scriptures; instead, he undertook to demonstrate from the structure and working of the universe, 'the mighty Volumes of Visible nature', from the starry firmament on high down to the lowliest insect, that the universe itself proclaimed its divine origin. There were no 'two cultures' in those days! The exceedingly learned James Ussher (1581–1656), Archbishop of Armagh, and the most learned Hebraist of his day, John Lightfoot (1602–75), had together calculated on the basis of mediaeval historical guess-work that the creation of the world had taken place in 4004 B.C., a date which began to be printed in the margins of the Authorized Version of the Bible (published by the university presses) from the year 1701 onwards right through the nineteenth century.

The firſt Age of the World: From the Creation to the Flood: This ſpace is called, ***Early in the morning, Mat.* 2.**

Hilar. in loc.

Ten Fathers before the Flood.

																						Adam hath *Cain* and *Abel*, and loſeth them both, *Gen.* 4. unhappy in his children, the greateſt earthly happineſs, that he may think of Heaven the more.
130	130																					*Seth* born in original ſin, *Gen.* 5. 2, 3. a holy man: and father of all men after the Flood, *Numb.*24.17. to ſhew all men born in that eſtate.
235	235	105																				*Enoſh* born: corruption in Religion by Idolatry begun, *Gen.* 4. 25. *Enoſh* therefore ſo named, *Sorrowful.*
325	325	195	90																			*Kainan* born: *A mourner* for the corruption of the times.
395	395	265	160	70																		*Mahalaleel* born: *A praiſer of the Lord.*
460	460	330	225	135	65																	*Jared* born: when there is ſtill *a deſcending* from evil to worſe.
622	622	492	387	297	227	162																*Enoch* born: and *Dedicated* to God: the ſeventh from *Adam, Jude* 14.
687	687	557	452	362	292	227	65															*Methuſhelah* born: his very name foretold the Flood. The leaſe of the world is only for his life.
874	874	744	639	549	479	414	252	187														*Lamech* born. A man *ſmitten* with grief for the preſent corruption and future puniſhment.
930	930	800	695	605	535	470	308	243	56													*Adam* dieth: having lived 1000. years within 70. Now 70 years a whole age, *Pſal.* 90. 10.
987		857	752	662	592	527	365	300	113	57												*Enoch* tranſlated: next after *Adams* death: mortality taught in that, immortality in this.
1042	*Enoch* the ſeventh from *Adam* in the holy line of *Seth*, prophecied againſt the wickedneſs that *Lamech* the ſeventh from *Adam* in the curſed line of *Cain* had brought in.	912	807	717	647	582	*Enoch* lived as many years as there be days in a year, viz. 365. and finiſhed his courſe like a Sun on earth.	355	168	112	55											*Seth* dieth.
1056			821	731	661	596		369	182	126	69	14										*Noah* born a comforter.
1140			905	815	745	680		453	266	210	153	98	84									*Enoſh* dieth.
1235				910	840	775		548	361	305	248	193	179	95								*Kainan* dieth.
1290					895	830		603	416	360	303	248	234	150	55							*Mahalaleel* dieth.
1422						962		735	548	492	435	380	366	282	187	132						*Jared* dieth.
1536								849	662	606	549	494	480	396	301	246	114					The CXX. years begin, *Gen.* 6. 3.
1556								869	682	626	569	514	500	416	321	266	134	20				*Japhet* born.
1558								871	684	628	571	516	502	418	323	268	136	22	2			*Sem* born.
1651								964	777	721	664	609	595	511	416	361	229	115	95	93		*Lamech* dieth.
1656								969		726	669	614	600	516	421	266	234	120	100	98	5	*Methuſhelah* dieth, and the Flood cometh.

Figure 1 From John Lightfoot, The Works of the Reverend and Learned John Lightfoot, *volume 1, 1804 (courtesy of the Evangelical Library).*

As far as the study of the historical sciences (as distinct from the natural sciences) was concerned, the Middle Ages did not end until the later years of the eighteenth century. Newton belongs to the twilight zone between mediaeval and modern man.

He had no inkling of the real character of the books of Daniel and Revelation (now known to be revolutionary tracts belonging to the second century B.C., and the end of the first century A.D. respectively) and he spent more time in his later years trying to decipher 'the number of the Beast' (Rev. 13: 18) than on his mathematical problems. It was assumed by the vast majority (including the clergy: there were no theological colleges in those days) that Ussher's dates were divinely inspired like the text of the pages on which they appeared.

The philosopher David Hume (1711–76) is nowadays respected as one of the most influential philosophers that Britain has ever produced. But in his own day (as he complained) his philosophical writings fell 'dead-born' from the press, whereas his *History of England* was widely acclaimed and read for over a century. It is the epitome of the eighteenth-century point of view. Hume states most explicitly the static view of history which prevailed throughout the Age of Reason: 'Mankind are so much the same, in all times and places, that history informs us of nothing new or strange. . . . Its chief use is only to discover the constant and universal principles of human nature. . . . Would you know the sentiments, inclinations and course of life of the Greeks and Romans? Study well the temper and actions of the French and English. . . .' Hume's perspective was still that of Marcus Aurelius: 'A man of forty years of age has seen all that has been or ever shall be.' The view of the ancients had changed not at all: 'That which has been is that which shall be . . . there is no new thing under the sun' (Ecclesiastes 1: 9). The hypothesis of the evolution of organic species could not have been conceived by anyone whose mind had been conditioned by the categories of thought of the Enlightenment period.

3.2 The development of the new historical method

The pioneers of the study of history as we know it today were already at work before the Age of Reason had ended. They were gradually developing the techniques of the literary and historical criticism of ancient records. Instead of simply copying down what ancient writers had said and repeating it, they started asking what lay behind the form of words which appeared in an ancient document, how that document had been put together from different sources, and whether in the light of what could be inferred about its composition the facts which it purported to relate could be considered to be true. The development of modern critical historical methods was begun by scholars at work upon the Bible. Sometimes these scholars were rationalists whose aim was to destroy the credibility of the Biblical history; one such was J. D. Michaelis who in 1750 published his *Introduction to the New Testament*. By 1753 a French Catholic, the physician Jean Astruc, had published (though anonymously) his analysis of the two sources of the Book of Genesis (nowadays familiar to every Biblical student as J and E). The new critical history was on its way, whether with the object of discrediting the notion of inspired Scripture or by way of elucidating its true and original meaning. It was at Göttingen about 1760 that the development of a genuinely historical sense, combined with critical acumen, was achieved: the names of J. C. Gatterer, J. C. Schlözer and J. S. Semler should be remembered as those of the men who opened the study of history in the sense in which the subject is taught in the universities of the world today. It has almost nothing in common with the propagandist history-writers of the *philosophe* school, who wrote history (as Hume told us of himself) with their feet up on the sofa. It was in 1787 that J. G. Eichhorn (who became a professor at Göttingen in the following year) during the course of his studies of Genesis invented the phrase 'the higher criticism' and established his claim to be the real founder of the study of historical documents in the modern sense (see J. Estlin Carpenter, *The Bible in the Nineteenth Century*, Longmans, 1903; Herbert Butterfield, *Man on his Past*, Cambridge University Press, 1955, especially pp. 26–58). The publication by the philosopher G. E. Lessing of his *Education of the Human Race* (1780) marks a turning-point in the development of human thought, one without which the biologists (or 'naturalists') of the nineteenth century would never have begun to look for the evidence in support of the hypothesis now 'in the air', the idea of evolution, an idea made familiar by the historians and Biblical scholars (usually the same people). Lessing taught that revelation (in Scripture and afterwards) meant *progressive* en-

Figure 2 G. E. Lessing (Mansell Collection).

lightenment. The Old Testament was only a preliminary stage in the process, suitable for the childhood of the human race. The New Testament marks a higher stage, but it will itself be superseded by the coming of a genuinely rational awareness of the divine purpose for mankind. Further generations will eventually realize all the high religious and moral potentialities of human nature.

In what way do these views of Lessing constitute a bridge between ideas characteristic of the eighteenth and nineteenth centuries?

Lessing is a man with one foot in eighteenth-century rationalism and the other having already a toe-hold in the nineteenth-century doctrine of inevitable progress. He died in 1781; the title of his posthumous essay (1784) is significant of the change which had already taken place in the attitudes of the more advanced scholars of the age: 'A New Hypothesis concerning the Evangelists considered simply as Human Historians.'

The latter expression is especially significant. Many of the deists and rationalists of the Enlightenment did indeed deny that the Gospels were divinely inspired, but they would have said not that the evangelists were historians but that they were writers of works of fiction. The conception of the Gospels (or indeed of other parts of the Bible) as works of history, which could be investigated by the methods of the new historical criticism, was an entirely original idea in the latter half of the eighteenth century. The revolution in historical method of the period from 1760 to 1860 was as important in the development of human thought and life as was the revolution in scientific method in the period between Copernicus and Newton. Thus Leopold von Ranke, professor of history at Berlin from 1825, helped to found the 'scientific' school of historians whose approach was soon applied to Christian origins and the Bible.

Of course, it must not be forgotten that other agents besides academic researches were at work in producing the distinctive 'climate of opinion' (to use a seventeenth-century phrase) which prevailed by the mid-nineteenth century. One of these which must be mentioned (but cannot be elaborated here) was the so-called Industrial Revolution, which had its own contribution to make towards belief in 'progress', the great Victorian dream, so rudely shattered in 1914. To many, of course, that revolution brought exploitation and misery, but to the well-to-do classes, who created the climate of opinion in the age of Darwin, it brought an era of comfort and prosperity such as had never been enjoyed by any earlier generation. Napoleon and Wellington moved their armies around in the same way and at the same speed as Alexander the Great or Julius Caesar had done. The age of steam introduced changes in the outward shape of things such as no previous age had ever seen. Those who would gain an impression of the changing scene as it appeared to people who lived in the earlier half of the nineteenth century should read the famous opening chapters of Macaulay's *History of England* (1848), with its vivid description of the rapidly changing scene. At no time before the twentieth century had so much outward or material 'progress' passed before the eyes of two or three generations. 'The history of England is emphatically the history of progress', said Macaulay in 1835.

Figure 3 Lord Macaulay, engraved by Sartain from an original by E. U. Eddis (Mansell Collection).

The long, slow process of evolution had been going on through countless centuries, but it did not make itself visible until the academic and the industrial revolutions of the hundred years or so after 1750 had combined to make people go and look for the evidence which would corroborate the impression, now widely spread abroad, that the earth and all the creatures on it were in a process of evolution. The recognition of this truth and the formulation of a theory of the evolution of animal and human species would have been impossible in any previous century, just as it would have been impossible for enquiring minds to have failed to perceive where the evidence was pointing in the nineteenth century. Intellectual history, like other types of history (political, economic, etc.), is not governed by immutable laws such as those imagined in Newtonian science in the world of nature, but it nevertheless has a logic of its own: that is to say, it is possible *after* the event to see why this or that development should have taken place when it did and not in any previous age. Darwin belongs to the nineteenth century just as inevitably as Newton belongs to the seventeenth; neither could have done what he did in any previous age. Intellectual history proceeds by the logic of its own development, but it is not governed by any laws which enable us to predict what the 'climate of opinion' will be like in the twenty-first century. The future is always unpredictable and full of surprises. This is one of the truths of history which men never seem to learn, and it might be said to illustrate the saying that the only truth we learn from history is that men learn nothing from history.

4 The Emergence of the New View of the Bible, 1760–1860

Figure 4 E. B. Pusey, by 'Ape', Vanity Fair *cartoon (Mansell Collection).*

The reaction to the 'higher criticism', when the news of it eventually crossed the German Ocean, was not unnaturally disturbing to ordinary clergymen and laymen in England. This was because of its association with the views of extreme rationalist propagandists like H. S. Reimarus (1694–1768) and H. E. G. Paulus (1761–1851), who sought to use the new historical approach to the Biblical literature in their assault upon traditional Christian doctrine. E. B. Pusey (1800–82), who later became the leader of the Oxford Movement, as a young Fellow of Oriel College, went to Göttingen and Berlin in order to study the new Biblical scholarship of the German critics. (He is said to have been one of the only two Oxford dons who could speak German at that time.) He also studied Semitic languages there; he became Professor of Hebrew of Oxford in 1828 and remained in that Chair for the rest of his life. In 1828 he published his *Historical Causes of the Rationalist Character lately predominant in the Theology of Germany* (Part II, 1830). Discouraged by its reception he later withdrew the publication and became prominent as the dominant figure of the Tractarian cause ('Puseyism'). What might have happened if Pusey had continued to expound the principles of Biblical historical scholarship instead of becoming the leader of the Oxford Movement is one of those 'ifs' of history which both fascinate and exasperate the enquiring mind; in history we cannot, as we can in the natural sciences, set up the experiment again and see how it would work if we varied one or other of the factors. The English Church in the middle years of the nineteenth century was preoccupied with the work of the Evangelical Revival on the one hand and with the issues raised by the Oxford Movement on the other. Theological scholarship at the universities was at a low ebb. The so-called 'higher criticism' was feared and misunderstood; the publication in Germany of D. F. Strauss' *Life of Jesus Critically Examined* (1835–6), which purported to show that the Gospels were second-century 'mythological' writings, aroused a storm of protest in Germany and later in England after the work had been brilliantly translated in 1846 by the youthful Mary Anne Evans, better known as George Eliot after she adopted that pseudonym in 1857 for the publication of her first novel. Paradoxically, but understandably, the new historical method, as developed by the original Biblical scholars, was immediately used by the rationalists, deists and assorted free-thinkers of the earlier nineteenth century in their polemic against traditional Christian doctrine, despite the fact that those who first devised and practised the higher criticism were devoutly Christian scholars seeking to find the truth in Holy Scripture and concluding that the Scriptural writings could be better understood if ancient Greek notions about inspiration were discarded. These scholars recognized that the Scriptures were written by men of their own time, who knew nothing at all about the Newtonian physics and were not concerned to convey geological or other kinds of scientific information to people of future generations. The truth contained in Scripture was true for men in every age, whether they held that the world was at the centre of the universe or in orbit round the sun; whether it was created in six days or 6,000 million years, and so on. Or, as it might be put in the light of the insights of perhaps the most profound religious thinker of the nineteenth century, a contemporary of Darwin, S. Kierkegaard (1813–55): the truth of the Bible is *existential*, not scientific; it concerns a man's existence as man, here and now, not any kind of truths about his biological origins or his physico-chemical make-up. Until the teaching of Kierkegaard had been assimilated, the two kinds of knowledge, the existential and the scientific, had never been clearly distinguished from one another. But, as so often happens in the history of thought, the teaching of Kierkegaard remained virtually neglected in his own age; its significance was not noticed until the twentieth century.

Figure 5 Søren Kierkegaard (Mary Evans Picture Library).

However, there was in England after about 1820 a group of men who were thoroughly alive to the issues raised by the newly discovered methods of historical study. It was probably their collective influence which prepared the rising generation for the inevitability of the triumph of the new historical methods now beginning to be accepted by Biblical scholars. This group included Thomas Arnold (1795–1842) of Rugby; Connop Thirlwall (1797–1875), Bishop of St David's; J. C. Hare (1795–1855) of Trinity College, Cambridge; H. H. Milman (1791–1868), Dean of St Paul's, and A. P. Stanley (1815–81), the secretary of the famous Royal Commission which enquired into the state of Oxford University (1850) and who subsequently became Dean of Westminster. This group was closely associated with their older contemporary S. T. Coleridge (1772–1834), whose acquaintance with German thought was unusual in those days. Indeed, the whole group was much influenced by the new and profound research of the new-style German historians, especially B. G. Niebuhr (1776–1831) from whom they learnt to write their own historical works (e.g. Milman's *History of the Jews*, 1829); Thirlwall in his youth actually ventured to translate a theological work of Schleiermacher, a popular 'advanced' theologian in Berlin around the turn of the century; later with J. C. Hare he translated Niebuhr's epoch-making *History of Rome*. The considerable influence of this group of 'Liberal Anglicans', as they have been called, was one of the chief factors in preparing the educated religious public for the storm which would inevitably come when the new historical approach to the Scriptures was at last made known to the ordinary clergy and laity of the country. The anxiety about 'the truth of the Bible' was already mounting when Darwin's *Origin of Species* was published in 1859.

Figure 6 Barthold Georg Niebuhr, from a sketch by J. Schnorr von Carolsfeld, engraved by Ruschewey (Mansell Collection).

In the light of the foregoing passages, would you agree with the statement in the second sentence of the next section that the so-called Darwinian controversy was in reality only a side-issue in a more important battle? Is the idea of 'science versus religion' a regular misconception?

Clearly one major issue at stake was the nature of Biblical revelation, and the extent to which it was right to draw scientific conclusions from the Bible. But it was more than the debate about 'accommodation' all over again. The recognition of a mechanism by which the Bible had been assembled had led some to suppose that their view of its status as the word of God must be altered. If, as Jowett said in *Essays and Reviews*, one should 'interpret the Scripture like any other book', it is pointless to seek for deeper insights than those which might be provided by other purely man-made works.

There is, however, an important parallel between this situation and that of cosmology. With the Copernican transformation what was really at issue was the *world picture*, or the mechanism of the universe. There was no *necessary* threat to the *world view* (Unit 2, pp. 56–7). In a similar way there was, and is, no *logical necessity* for a change of view regarding Biblical origins to be accompanied by a devaluation of its status as the Word of God. The debate on this continues, and in the early nineteenth century many saw the point just made, although, as we have seen, others actively promoted Biblical scholarship in the hope that the Bible would be generally discredited.

If this was the real nature of the controversy, then it would seem that here is another case of the 'conflict thesis' of science and religion having been considerably overstated. Consistent with this assessment is the fact that the Cambridge group of 'Liberal

Anglicans' included such major scientific figures as Herschel, Babbage, Henslow and Sedgwick. One recent scholar has commented of this group that it 'represents the professionalization and internationalization not only of English science but also of English scholarship and religious thought', each member seeing 'his work not as an isolated unit, but as part of an intellectual totality' (W. F. Cannon).

5 The Crisis of Belief in the Later Nineteenth Century

It is almost impossibly difficult for us, little more than a century later, to comprehend the feverish anxiety which gripped many ordinary Christian folk when the crisis of belief concerning the nature of Biblical authority had reached its climax about the year 1860. Though popular imagination has fixed upon Darwin's *Origin of Species* as the cause of the excitement, it was only a side-issue in the conflict. It caused neither surprise nor bewilderment to the educated minority who were conversant with the attitude of those outstanding Anglican churchmen of the Arnold–Thirlwall group. They had already abandoned the traditionalist doctrine of Scripture and were teaching the new view that God revealed His truth gradually as men were able to understand it, finding it and assimilating it for themselves under the inspiration of the Holy Spirit, whose guidance was accommodated to the limitations of human understanding in any given age. Thus, the Bible itself showed that there had been an evolution from the first dawning of the human religious consciousness to its full development in a St Paul or St John. The Biblical books had been written just like other human writings; the difference lay in the subject-matter, the *kind* of truth with which they dealt. The Biblical writers were not mere 'pens in the hand of the Holy Spirit'; they were men, inspired indeed, but not omniscient, who could not know truths which had been discovered centuries after their time.

Figure 7 John William Colenso, Bishop of Natal (Mansell Collection).

It is not surprising that most ordinary people were offended by the new way of looking at the Bible in the light of modern knowledge; and indeed some of those who published the results of their researches exercised little discrimination in their methods of educating the public. Amongst these, for example, was J. W. Colenso (1814–83), who became Bishop of Natal in 1853. In 1861 he created a furore by questioning the doctrine of eternal punishment in his *Commentary on the Epistle to the Romans*, but this was nothing compared with the controversy which broke out when he began to publish his *The Pentateuch and the Book of Joshua Critically Examined* (in parts between 1862 and 1879). The Bishop of Cape Town deposed him in 1863 but the Judicial Committee of the Privy Council gave sentence in his favour in 1865. The controversy dragged on even after his death and the schism in the South African Church was not finally healed until 1910. Colenso gave offence by his liberal attitude generally (e.g. in permitting polygamists to retain their several wives after baptism); the mid nineteenth century was a sternly illiberal age.

There were, however, those who believed that freedom in religious enquiry should be maintained. In *Essays and Reviews* (1860) seven contributors published their scholarly thoughts upon certain religious questions of the day; the best known of them are Benjamin Jowett (Master of Balliol), Mark Pattison (Rector of Lincoln College), and Frederick Temple (at that time Headmaster of Rugby). The essays represented a kind of thinking aloud on the questions that had arisen in the new theological climate of the day. They did not directly discuss the theory of biological evolution which had been publicized by the appearance in the previous year of Darwin's *Origin of Species*. The essayists were not particularly interested in Darwin's theory of natural selection, namely, that only the species or members of a species which are best adapted for their biological purpose will survive in the struggle for existence. It was the question of the relationship between Biblical authority and scientific progress which was foremost in the minds of the essayists. C. W. Goodwin argued that a literal interpretation of Genesis was at variance with the truths discovered by the geologists, a view generally accepted by educated Christians in 1860; and by that date it was not very shocking to hear Baden Powell repeating that the miracles were not evidence for the truth of Christianity. The essay which most seriously disturbed religious leaders and churchmen was significantly that of Benjamin Jowett on 'The Interpretation of Scripture'.

Here is a short excerpt from this essay:

> Scripture has one meaning, the meaning which it had to the mind of the prophet or evangelist who first uttered or wrote, to the hearers or readers who first received it. Another view may be easier or more familiar to us, seeming to receive a light and interest from the circumstances of our own age. But such accommodation of the text must be laid aside by the interpreter whose business is to place himself as nearly as possible in the position of the sacred writer.

Figure 8 Benjamin Jowett, by 'Spy', Vanity Fair *cartoon (Mansell Collection).*

How far does Jowett's advice conflict with the seventeenth-century 'principle of accommodation' (see Unit 2, p. 71)?

Although Jowett attacks 'accommodation' he uses the word in a different sense: the reading into a passage of our own modern experience. He is in fact (though perhaps unconsciously) closely following Calvin and others in accepting the importance of the original sense.

The new historical approach to the study of the Biblical literature was less familiar to the clergymen who signed the protest against *Essays and Reviews*. Unlike the essayists, the country clergy in the 1860s had not assimilated the teaching of the Old Testament scholars that the Bible itself bore decisive evidence that men's understanding of truth had evolved from childish beginnings to high levels of sophisticated reasoning and deep awareness of divine truth. The new historical scholarship which had altered the whole perspective of the Age of Reason and had made Darwinian science possible belonged, of course, to the intellectual furniture of the essayists' minds. They were content to let the scientists themselves wrangle over the Darwinian hypothesis concerning natural selection. (Reading good biographies of the protagonists in the *Essays and Reviews* controversy is one of the best ways of understanding what the situation felt like in days so very far removed in outlook, if not in date, from our own. It happens that excellent biographies of two of the protagonists have appeared in recent years: Geoffrey Faber on *Jowett*, Faber and Faber, 1958, and V. H. H. Green, *Oxford Common Room* (on Mark Pattison), Edward Arnold, 1957).

Essays and Reviews would seem to any curious reader who picked it up today a rather dull and commonplace volume. The essay of Jowett, as has been noted, caused most debate and thereafter he abandoned writing on theological themes. On publication the volume received an unfavourable review in the (Church) *Guardian* but attracted little attention until Bishop Wilberforce launched one of his characteristically intemperate attacks upon it in the *Quarterly*. Samuel Wilberforce was one of those vociferous and dynamic personalities whose impressive oratory and restless energy are not regulated by equivalent intellectual ability or the capacity of balanced judgement. He was the third son of the respected William Wilberforce of Hull, the philanthropist and dedicated advocate of the abolition of the slave trade. Samuel was Bishop of Oxford from 1845 till 1869. He had indeed many excellent qualities and worked as a reformer in many good causes. But his excursions into theological controversy were disastrous. He so powerfully defended the causes which he took up that they collapsed under the weight of his advocacy. The violence of his denunciation in the *Quarterly* and his subsequent campaign against *Essays and Reviews* secured the condemnation of the volume by the Bishops in a form which the Archbishop of Canterbury issued as an encyclical. Legal action was taken against two of the contri-

Figure 9 Bishop Samuel Wilberforce (Mansell Collection).

butors, whose names are forgotten today (H. B. Wilson and R. Williams), but happily the Judicial Committee of the Privy Council quashed the sentence of deprivation for one year which had been passed upon them. (The cause of liberalism in the later nineteenth century owed a debt to the sensible laymen of the Privy Council.) A declaration drawn up by way of protest against the spread of liberal views was signed by 10,906 clergymen, who affirmed their belief in the verbal inspiration of Scripture and the doctrine of eternal punishment. In 1864 the volume was given formal synodical condemnation. The value of uninformed reactions of this kind may be measured by the fact that one of the contributors (Temple) subsequently became Bishop of Exeter (despite fierce protests) in 1869 and Archbishop of Canterbury in 1897. It is generally held by historians of the nineteenth-century Church that it was largely through the effective power of laymen ('the Establishment') that the Church of England was able to build up a learned leadership amongst the ranks of the higher clergy (e.g. Lightfoot and Westcott, world-famous New Testament scholars, who were successively Bishops of Durham).

Figure 10 Frederick Temple, photograph by Samuel A. Walker (Mary Evans Picture Library).

6 Some Difficulties in the Reception of the Darwinian Hypothesis

It is against this background that both the popular and the clerical reaction to the Darwinian hypothesis should be seen, when the *Origin of Species* was published in 1859.

In the light of previous units can you suggest what difficulties attended the reception of *Darwin's* views? See also *Anthology II*, pp. 282–7.

In brief these would include:

1 Difficulties in communication.
2 Problems of suffering.
3 Argument from design.
4 Problems of mechanism.
5 The status of man.
6 Social Darwinism.

They are briefly considered in the following pages.

6.1 Difficulties in communication

The occasion which drew the views of Darwin to the attention of the public was the meeting of the British Association (founded in 1831) in 1860. Bishop Samuel Wilberforce was not the man to miss this opportunity of denouncing 'Darwinism', and, as was his wont, he did so in unmeasured terms. His argument was that 'the principle of natural selection is absolutely incompatible with the word of God'. The 'defence' of the traditional orthodoxy by Wilberforce caused great embarrassment to the thoughtful churchmen who had long ago accepted that 'the word of God' was not to be equated with the verbal inspirationist view of it. T. H. Huxley (1825–95), the self-appointed spokesman for 'science', was able without difficulty to dispose of Wilberforce's bluster in the debate. The result of this unedifying scene was the arousing of public excitement over the issue, which now became a *cause célèbre*. Everyone took sides, including the leading politicians (Gladstone, and Disraeli – 'on the side of the angels'). No one was more surprised than Darwin himself at the public sensation which was created. He had not expected anything of the kind and had told his publishers that 1,250 copies of the *Origin of Species* would suffice. All at once he found himself world-famous and involved in a controversy which he did not enjoy and which he had not expected. His work when translated into other languages caused similar excitements abroad, especially in Germany. Unfortunately the echoes of the controversy, still couched in the language and based on the assumptions of the mid-nineteenth century, are about all that the British public knows of the issues which were then being debated. The nearest parallel in our own times to the scandal (in the proper sense of the word) of the 1860 controversy would be that created by the book of J. A. T. Robinson, then Bishop of Woolwich, *Honest to God* (1963). In it the Bishop attempted to explain the standpoint of such eminent theologians as Bultmann, Barth and Tillich, all born in the 1880s, and whose works had long been translated into English and other languages. A sensational article in *The Observer* resulted in the fact that the book was sold out in a day and many thousands of copies were subsequently printed. Such is the time-lag between popular awareness of religious and scientific discussion and the actual state of contemporary thinking by theologians and scientists themselves.

For thoughtful churchmen (typified, for example, by those educated at, say, Rugby School in the period of the headmasterships of Arnold and Temple) the idea of evolution, whether of the universe or of organic life, was no novelty; but there was in those days, before the coming of compulsory education for all (1870), a wide gap between the educated minority and the great majority of the population, who did not even have the 'popular' press to stimulate them, because they could not read. The educated minority would, however, include a number of 'working' men, such as those who had attended the lectures at F. D. Maurice's Working Men's College founded in London in 1854 or had come into contact with the educational activities of Maurice's friends, the 'Christian Socialists', such as J. M. F. Ludlow and Charles Kingsley; but these would represent only a very small proportion of the working classes of the mid-nineteenth century. (It was the 1870 Education Act which made the publication of *The Daily Mail* a possibility in 1896; before about that date newspapers had been produced for and read only by the literate public.)

6.2 Problems of suffering

The liberal Anglican clergy were quite content to let the scientists debate the scientific question whether in fact the distinctive Darwinian thesis was sound, namely, that it was the survival of the fittest in the struggle for existence which was the main factor in the development of biological species. Darwin's stress upon this factor certainly posed the ancient problem of evil in a new form, how a God of love could have created a world in which there was so much cruelty and carnage. It was a problem which already haunted the minds of thoughtful people in the nineteenth century, long before Darwin gave it prominence in his doctrine of the struggle for existence and the survival of the fittest. Tennyson in 1850 had already expressed the horror of 'nature red in tooth and claw' (*In Memoriam*, lvi; cf. lv) and written of 'Nature . . . so careful of the type . . . so careless of the single life'.

6.3 Argument from design

The Darwinian hypothesis also at first sight seemed to create a difficulty for Christian apologetic as expressed in the so-called argument from design (the third of the three traditional 'proofs' of the existence of God, the teleological argument). William Paley's *Natural Theology* (1802), a book widely read and admired throughout the nineteenth century, had argued convincingly to the existence of God from the evidence of design throughout the whole creation, the perfection of a butterfly's wing or of a human eye: the fact of such intricate design throughout nature implied the existence of a Designer. On the older scientific view, the Newtonian-mechanistic theory, the argument was convincing: 'Paley's Watch' became famous in Christian apologetic because of the strength of the good Archdeacon's argument that the existence of a machine logically implied the existence of a Designer. But the new Darwinian science replaced the category of mechanism by that of organism. The butterfly's wing or the human eye had evolved from rudimentary beginnings in the process of evolution: the organisms which had succeeded in adapting themselves to their environment by developing better wings, eyes, brains, etc., in the struggle for existence were the ones which had survived. No theory of design was necessary. All that was needed was not the concept of God as Paley's great Watchmaker but closer observation of the evolutionary process itself, which suggested not a Creator-God but a powerful Life Force which did not pursue a conscious end but was rather like a giant who walked in his sleep. The rationalists were quick to seize upon the Darwinian theory in their polemic against Christianity. The continuing argument was as fierce as it was tedious and endless. Surely, it was said in reply to the rationalists, the perfect beauty of the wing of, say, a peacock butterfly was hardly developed just in order to persuade these lowly creatures to breed. To account for the beauty of nature by means of utilitarian biological urges is to step out of the world of science into the underworld of rationalistic philosophy.

'There cannot be a design without a designer' (Paley, *Natural Theology*, 16; also *Genesis and Geology* (SET BOOK), p. 36). This sentence is logically impeccable, but it does not prove what Paley wishes to assert. The prior question is: Is there evidence of design in nature? This is not a scientific but a philosophical question. Some people just feel that there is no design at all: things just happen to be like that. They need no more 'explaining' than any other given 'fact'. I enjoy a beautiful sunset or Mozart in the same way that I enjoy a warm fire on a cold day: it affects my comfort, whereas another person might remain unaffected by the sunset (or by Mozart) because his make-up is different from mine. (Cf., for example, the standpoint of Antony Flew in *God and Philosophy*, Hutchinson, 1966.) Other people find themselves 'compelled' to believe in the existence of 'design' and that it implies (logically) the existence of a Designer (though, of course, he may be loving, indifferent, doing his feeble best, or simply a doodler). Would you agree that the fact of biological evolution in a Darwinian or any other sense has no necessary bearing upon the question of Design-Designer/Chance-Meaninglessness? (Another way of speaking about the same theme would be to ask about 'meaning' instead of 'design'.) Natural science does not provide answers to philosophical questions. The argument, of course, continues, and it is not our concern here. We should, however, take note of it because of its importance in the controversies of the post-Darwinian nineteenth century. See also Units 9–10 and chapter 9 in *Anthology II*.

6.4 Problems of mechanism

Another difficulty that arose at an early stage is still with us.

No one would wish to base any argument upon our ignorance, but it would be foolish to forget that a considerable number of questions concerning biological evolution are still unanswered; it may be that when (or if) they are answered, our problems would appear in a new perspective. An example of an unanswered biological question might be: How did insects 'evolve' their wings? Birds, bats, pterodactyls, etc., developed the ability to fly by the modification of their limbs into wings which enabled them to survive in the struggle for existence. But insects' wings (which enabled them to fly about 400 million years ago) are independent outgrowths of the thorax. What could have caused wings to evolve, if insects were not 'created' with them in the first place? (See Sir Vincent Wigglesworth's discussion of the question in the magazine *Nature*, 16 November 1973.) If we knew the answers to many more questions of this kind (but much more complex than this one) we might be able, not indeed to answer philosophical and theological questions, but at least to formulate our questions more adequately.

6.5 The status of man

This was a basic issue in the debate between Wilberforce and Huxley. It raises some perennial questions. For example, how far is it true to say that Darwin's 'offence' arose from the challenge to human dignity and pride by his theory of man's descent from the higher apes? (Cf. the debate between Wilberforce and T. H. Huxley.) Would it in any way affect a Biblical-Christian doctrine of human nature, if Darwin's account of man's descent should be held to be probable (cf. Psalms 48: 8; 49: 20)? Was man a special creation or not?

6.6 Social Darwinism

See the remainder of chapter 14 in *Anthology II.* The extension of what purports to be a purely scientific theory to one which is in effect a social philosophy is fraught with difficulties. Again, we can merely draw attention to the problems. Thus, is it possible to base ethical theories upon scientific investigations? (Cf. the 'evolutionary ethics' of T. H. Huxley.) Are such theories bound up with Victorian optimism about progress ('let the ape and tiger die')? If Darwin's view of the descent of man were held to be scientifically acceptable, would it in any way affect the question whether man held a unique place in nature (or, in religious language, the Creation)?

7 The Basic Question for the Victorian Church

The basic question for the Church in the nineteenth century, however, concerned 'the truth of the Bible'. Before Darwin was born and after his death the question was being debated; it was not created by the publication of his theory of biological evolution, nor was it in any important way affected by the debate about evolution.

For a long time before Darwin's work was published a number of eminent, learned and devoutly Christian scholars, students of 'the new history', had already abandoned the older view of the verbal inspiration of the Scriptures. The three greatest Biblical scholars in England in the nineteenth century, of whom Lord Acton said that they were the equals of any German or Frenchman, were the Cambridge trio, B. F. Westcott (1825–1901), J. B. Lightfoot (1828–89) and F. J. A. Hort (1828–92). They laid the foundations of English New Testament scholarship and guided the course of the theological revolution which was taking place in their days and was shaping the future of theology in the English-speaking world. (The justly renowned critical edition of the Greek New Testament published in 1881 was the result of thirty years' close collaboration between Westcott and Hort. It is worth pointing out that Hort was at the same time acting as an examiner in the Natural Science Tripos at Cambridge.)

Wescott succeeded Lightfoot as Bishop of Durham in 1889. Despite the views of most of the leaders of thought in the Victorian Church, it is hardly surprising that there should have been vigorous opposition to them, because the change of outlook required to appreciate the latter was far greater than the change required in the transition from the Ptolemaic to the Copernican astronomy.

Why would such a change of outlook be so difficult to achieve?

It was harder to accept because it touched men's deepest convictions at the existential level: it involved not merely external matters, such as whether the sun went round the earth, but the very nature of belief itself.

Many conservative preachers, whether evangelical or tractarian, inveighed against the new liberalism. The most widely quoted reassertion of the traditional position is that of an Oxford theologian, afterwards Dean of Chichester, J. W. Burgon (1813–88), who in 1861 asserted the traditionalist position from the university pulpit:

> The Bible is none other than the voice of Him that sitteth on the throne. Every book of it, every chapter of it, every verse of it, every word of it, every syllable of it (where are we to stop?), every letter of it, is the direct utterance of the Most High. The Bible is none other than the Word of God, not some part of it more, some part of it less, but all alike the utterance of Him who sitteth upon the throne, faultless, unerring, supreme.

The opposite point of view was expressed in a letter written by Frederick Temple, then Headmaster of Rugby, to the Bishop of London, A. C. Tait, who became Archbishop of Canterbury in 1868:

> Many years ago you urged on us from the University pulpit to undertake the critical study of the Bible. You said it was a dangerous study, but indispensable. . . . Such a study, so full of difficulties, imperatively demands freedom for its condition. To tell a man to study, and yet bid him, under heavy penalties, come to the same conclusions with those who have not studied, is to mock him. If the conclusions are prescribed, the study is precluded.

Tait had, in fact, joined the other bishops in publicly deprecating *Essays and Reviews*.

It is now apparent why it was suggested above that the publication of Darwin's *Origin of Species* was little more than a skirmish in the serious struggle between those who wished to restrict scholarly research in the Biblical field and those who were actually engaged in it or supporters of it. The real issue was not between Genesis and Geology, or between Darwin and the Bible, but between the traditional and the critical approaches to biblical-historical study and interpretation.

Can you suggest a specific example of such a difference?

Perhaps the best known relates to multiple authorship of Biblical books. Thus, over a period of about a hundred years before 1859, the opinion was growing that Genesis was a composite work, not written by Moses but compiled by an editor or editors from several sources which themselves, as we now have them, were written down long after the time of Moses. One slightly later scholar, Wellhausen, developed this theme so thoroughly that he has been hailed as a Darwin of Biblical studies! And, of course, the debate continues.

But, of course, the popularization of Darwin's thesis drew the attention of thousands who knew nothing about the progress of Biblical and historical scholarship to a matter of which they were previously unaware. The shock was considerable for many simple believers, amongst whom must be reckoned a large number of the clergy. One good thing which Bishop Samuel Wilberforce did was to found Cuddesdon Theological College (1854); before that time the clergy generally had received no specialized theological training at all. The Honours School of Theology at Oxford and the Theological Tripos at Cambridge were not founded until the 1870s. Most of the clergy had received a good but old-fashioned classical education, which by itself did little to open their minds to the new theological studies. Hence their inability to cope with the intellectual pressure of works like *Essays and Reviews*.

How far is it true that Darwin's work profoundly affected theology?

The idea that the publication of Darwin's *Origin of Species* was decisive in the history of English theological thought is part of the mythology of nineteenth-century rationalism. The shock which it gave to an ill-educated and conventionally minded church-going public was indeed unpleasant for the many who disliked new ideas. But the effective leaders of religious thought in England did not view the matter as in any way decisive for faith or unbelief. They gave to Darwin the honour which was his due in the development of the modern scientific world view. Darwin died in 1882, a year after the death of A. P. Stanley, who had been Dean of Westminster from 1864. Darwin became more inclined to agnosticism in his later years, and he sadly acknowledged that his capacity for musical appreciation had atrophied from lack of use. He was buried in Westminster Abbey, a few feet away from the grave of Isaac Newton. Karl Marx, who had also utilized the characteristically Victorian concept of the struggle for existence in another field of investigation, died the following year and was buried in Kensal Green cemetery.

Darwin is important in the history of science in his own right, whatever the scientific verdict upon his theory of natural selection might be. He himself was reticent about expressing an opinion in the controversy which arose as between different schools of churchmen on the one hand or between Christians and rationalists on the other. Unlike Sir Charles Lyell at an earlier date, Darwin did not feel it necessary in 1859 to discuss the relation of his scientific discoveries and hypotheses to the Scriptural revelation. Lyell dissented from the common attempts to accommodate the age of the earth to the creation-stories of Genesis by means of the by then familiar hypothesis that the six days of the Creation really referred to six separate creative acts. In his *Geological Evidences of the Antiquity of Man* (1863) he expressed no dissatisfaction with Darwin's recently published conclusions, but he never expressed whole-hearted agreement with them. Like Darwin, Lyell was buried in Westminster Abbey. It is inconceivable that Lyell and Darwin, the two leading scientists of the day, should have been unfamiliar with the work of the liberal Biblical scholars; they can hardly have thought of themselves as having 'disproved' the Bible. They must have been aware that many leading scholars in the Church held that even the Scriptures themselves had 'evolved' over long ages. But they doubtless felt that no useful purpose was to be served by their becoming publicly involved in religious polemics in an age of extremist, even fanatical, religious controversy.

Figure 11 Sir Charles Lyell, engraved by C. H. Jeens (Mansell Collection).

T. H. Huxley, who was rather more than a dilettante publicist, having been a serious contributor to zoological scientific research, became the spokesman of 'advanced thought' in the later nineteenth century. He is credited with having coined the word 'agnosticism'. He believed that the Bible should be the basis of religious education in schools (after the 1870 Act), but he himself became more and more hostile towards Christianity. His most important and representative publication was his Romanes Lecture entitled *Evolution and Ethics* (1893).

It would appear that the top scientists and the top Biblical scholars and theologians deliberately stood outside the arena of the conflict over 'Darwinism' and Scriptural inspiration. They understood one another very well, met in the same clubs and societies and read one another's works: there were not 'two cultures' in the 1860s, and an intelligent man (a Jowett, a Hort, a C. L. Dodgson—alias Lewis Carroll) could have a fairly shrewd notion about what was being said at the frontiers of research in most of the spheres of human knowledge. In a scholarly way eminent theologians disagreed with one another from the university pulpit in Oxford or in their published works. Frederick Denison Maurice (who in 1853 had resigned his professorship at King's College, London, because of his rejection of the doctrine of eternal punishment) in his book *What is Revelation?* (1858) strongly attacked Canon H. L. Mansel's *Limits of Religious Thought* (the Bampton Lectures for 1858), which like other works of this period stressed that, although in secular historical studies ('profane history') immense advances had been made by the science of historical criticism, such methods could not be applied to 'sacred history' (i.e. Genesis, etc.), since the historical record concerning the beginning of the creation had been handed down by inspired men within five generations—Adam to Moses—and was therefore 'a history absolutely and in every respect true'. The dialogue between the conservatives and the liberals went on within the Anglican communion and the Free Churches for the rest of the nineteenth century, by which date it may be said that the issue had been settled in favour of the new historical approach to the Bible. In the Protestant churches there has always been a vocal minority[1] who cling to a view not so far removed from the theory of the mechanical dictation of Scripture.

The majority verdict of scholarship at the end of the nineteenth century is best represented by W. Sanday's *Inspiration*, the Bampton Lectures for 1893. Sanday, an immensely learned Oxford professor from 1882 until his death in 1920, taught the doctrine of 'progressive revelation', that God revealed his truth gradually to men as they slowly advanced in natural and spiritual knowledge, until at last in the fullness of time the truth could find its highest expression in Christ. Sanday's persuasive advocacy of modern methods of Biblical study was certainly a powerful influence in bringing the clergy and ministers of the churches to accept the new historical understanding of the Scriptural revelation. In the Roman Catholic Church the acceptance of the new critical methods was delayed as a result of the conservative reaction against 'modernism' until the publication of the papal encyclical *Divino afflante Spiritu* (1943).

8 A Concluding Question

The preceding section has raised a large question, which perhaps cannot be answered definitively but which is worthy of continuing meditation. It might help you to understand the depth of the question if you tried to write down your own reflections upon it. It is generally agreed that every age has to re-write for itself its own history of the preceding ages. This is because the historian always looks through the spectacles of his own age: the significance of (say) what happened in 1776 or 1789 looks very different two hundred years afterwards from how it appeared to writers only fifty or so years afterwards. Does not the history of science show a similar need of being re-written in every age? Are not scientists, if not to the same degree as historians, conditioned by the categories of their own age? Is purely objective judgement possible, except perhaps in the purest of pure sciences, viz. mathematics? Are not *generalizations* in the sphere of the biological, psychological and social sciences conditioned, at least to some extent, by the 'climate of opinion' of the age in which they were born? Does not nineteenth-century optimism colour the judgement of Darwin as it does that of Paley? Paley sees a vision of nature at peace with itself and enjoying its happiness: 'It is a happy world after all'; 'the world was made with a benevolent design' (on The Goodness of the Deity). Darwin says little about a Deity, but when (in the penultimate paragraph of the *Origin of Species*) he draws to his conclusion, he expresses the optimism of the Victorian age. Looking back over the long period of development from 'long before the Silurian epoch', he remarks: 'We may look with some confidence to a secure future of equally appreciable length. And as natural selection works solely by and for the good of each being, all corporeal and mental endowments will tend to progress towards perfection.' (This, it will be remembered, is the quotation with which we began.) The 'climate of opinion' in every age seems to contain its own peculiar

[1] Such people have often been called Fundamentalists but it may be doubted if this is the appropriate term. It seems to have been first used in 1920 to describe certain American Christians opposed to the removal of supernatural elements of Christianity as advocated by certain 'progressive' thinkers at the time. Some ten years earlier a series of twelve paperback books was issued with the general title 'the Fundamentals'; these re-emphasized certain traditionally held doctrines of the Church particularly those associated with the person of Christ. In neither of these cases was the primary thrust against objective Biblical scholarship and certainly 'The Mechanical Dictation Theory' was no part of their thinking. Since that day, however, the word Fundamentalist has acquired a pejorative meaning, relating to the position of extreme ultra-conservatism that had been discredited in the nineteenth century if not before.

infection, from which neither historians nor scientists are immune, nor yet philosophers or theologians. If only we could know what the historians of the later part of the twenty-first century will be saying about the ways in which opinion is being conditioned for us now in the 1970s! What then, finally, is the value of the study of the history of ideas?

Acknowledgements

Grateful acknowledgement is made to the following sources for illustrations used in this unit:

Figure 1: courtesy of the Evangelical Library; *Figures 2–4, 6–9 and 11:* Mansell Collection; *Figures 5 and 10:* Mary Evans Picture Library.

Science and Belief: from Copernicus to Darwin

The 'Conflict Thesis' and Cosmology

Towards a Mechanistic Philosophy

Scientific Progress and Religious Dissent

New Interactions between Theology and Natural Science

The Crisis of Evolution

The New Outlook for Science